TIME REVERSIBILITY, COMPUTER SIMULATION, AND CHAOS

ADVANCED SERIES IN NONLINEAR DYNAMICS

Editor-in-Chief: R. S. MacKay *(Cambridge)*

ADVANCED SERIES IN
NONLINEAR DYNAMICS
VOLUME 13

TIME REVERSIBILITY, COMPUTER SIMULATION, AND CHAOS

William Graham Hoover

University of California at Davis/Livermore

World Scientific
Singapore • New Jersey • London • Hong Kong

Published by

World Scientific Publishing Co. Pte. Ltd.

P O Box 128, Farrer Road, Singapore 912805

USA office: Suite 1B, 1060 Main Street, River Edge, NJ 07661

UK office: 57 Shelton Street, Covent Garden, London WC2H 9HE

British Library Cataloguing-in-Publication Data
A catalogue record for this book is available from the British Library.

First published 1999
Reprinted 2001

ISBN 981-02-4073-2

This book is printed on acid-free paper.

Printed in Singapore by Uto-Print

Frontispiece and Dedication

Josef Loschmidt (1821-1895 ⟷ 1895-1821) taught at the University of Vienna for a quarter century. He was a good friend of Boltzmann, an ingenious investigator, and a staunch defender of the atomistic view. He is best known for counting atoms, for determining molecular structures (including "Kekule's", for benzene), and for posing the fundamental reversibility paradox which concerns us here. For more information concerning this interesting man, see the reference to Fleischhacker and Schönfeld, listed in the Alphabetical Bibliography at the end of this book.

for Carol

Here, with a loaf of bread beneath the bough—
A flask of wine, a book of verse, and thou
beside me, singing in the wilderness;
and wilderness is paradise enow.

Omar Khayyám, as translated by Edward FitzGerald

Preface

Today a small army of physicists, chemists, mathematicians, and engineers has joined forces for a renewed attack on a classic problem, the "reversibility paradox". The paradox is simply stated: "How can the *irreversible* Second Law of Thermodynamics be compatible with, and result from, an underlying *time-reversible* mechanics?" Building on the ideas of van der Waals and Maxwell, Boltzmann provided the classic nineteenth-century resolution of the paradox by using a probabilistic analysis of dilute-gas collisions. Here I bring Boltzmann's classic analysis up to date by adopting modern tools. This approach augments and generalizes Boltzmann's statistical understanding. The new interpretive tools are (i) Linear-response theory, a consequence of Gibbs' statistical mechanics, (ii) Chaos theory and (iii) the Fractal geometry to which it leads, and (iv) Computers, which make possible the simulations and analyses which were not available to Boltzmann. As the available tools change, so do the targets and the points of view. Philosophers interested in the reversibility paradox have provided some insight too. We will seek correlations with their work.

The present book describes both the scientific and the philosophical work from the perspective of computer simulation, emphasizing my own *thermo*mechanical approach to resolving the reversibility paradox by analyzing the consequences of time-reversible thermostats. Computer simulation has made it possible to probe and characterize time reversibility from a variety of directions. "Chaos theory" or "nonlinear dynamics" has supplied a vocabulary detailing a useful set of concepts, which allows for a fuller explanation of irreversibility than was available to Boltzmann or even that provided by the linear response theory of Green, Kubo, and Onsager.

Throughout my own research career I have spent countless hours reread-ing the fruits of others' work. This rereading has been made necessary by the lack of clear example problems illustrating the meanings of the concepts used in these works. While this lack was understandable in the precomputer era of Boltzmann, Krylov, and Zubarev, it is inexcusable today. Through-out the book I emphasize the clear illustration of fundamental concepts with simple example problems, suited to desktop computation. I have also clarified the concepts by including a glossary of technical terms from the specialized fields which are combined here to focus on a common theme.

I am personally quite satisfied with the modern resolution of the re-versibility paradox, as presented here. I see thermodynamic irreversibility as an inevitable, understandable outcome of an underlying time-reversible dynamics. This understanding is predicated on accepting that *no* dynamics is a *perfect* replica of nature. I cannot imagine any completely comprehen-sive "unified theory", able to include all of nature, together with its ob-servers. My goal, in this book, is more modest. I wholeheartedly embrace *classical* mechanics as the most useful basis for an understanding of the physical world on the length and time scales relevant to us humans. By generalizing classical mechanics, to include temperature and thermostated "thermal boundaries", we obtain "thermomechanics". This discipline will be our main model for the exploration and explanation of the links between time-reversible micromechanics, macroscopic irreversible thermodynamics, computer simulation, modern chaos theory, and fractal geometry.

The book begins with a discussion contrasting the idealized determin-istic reversibility of basic physics with the pragmatic unpredictable irre-versibility of what we call "real life" or "nature". The chaotic complexity discovered by Maxwell, Boltzmann, and Poincaré suggests that the unpre-dictability of life is intrinsic. This view is quite consistent with Gödel's undecidability proof, as well as with our quite evident ability to affect the future by exercising "free will".

Computational models and simple thermomechanical simulations based on them are discussed and illustrated throughout the book. The simulations provide a reliable means to assimilate complex concepts through worked-out examples. Such analyses, from the point of view of dynamical systems, are applied to simple two-dimensional maps and higher-dimensional dy-namical systems, as well as to many-body examples from nonequilibrium molecular dynamics and to chaotic irreversible flows from finite-difference, finite-element, and particle-based continuum simulations. Two necessary concepts from dynamical systems theory—fractal distributions and Lyapunov instability—are fundamental to interpreting the results of the computational approach.

Undergraduate-level physics, calculus, ordinary differential equations, and a taste for computation are sufficient background for a full appreciation of the book. For nearly twenty years the Academy of Applied Science (Concord, New Hampshire) has sponsored the summer work of bright high-school seniors in the University of California's Davis Campus' Department of Applied Science at Livermore. The example problems worked out in the book are representative of the summer projects to which these students have contributed. The book is intended to appeal to advanced undergraduate as well as to graduate students, and to research workers. I fervently hope that the generous assortment of examples that I have worked out in the text will stimulate readers to explore and enjoy the rich and fruitful field of study which links fundamental reversible laws of physics to the irreversibility which surrounds us all. I have chosen mainly one- and two-dimensional examples in order to permit me to convey ideas with simple *pictures*. I stress here that the *ideas* so illustrated are not essentially different in three space dimensions.

To summarize the view I have reached, as the result of a decade of research, the Second Law of Thermodynamics is most simply described as a ubiquitous time-symmetry breaking which invariably accompanies the dynamics of a sufficiently chaotic system connected to its environment. Now it is certainly true that the "chaos" and symmetry breaking found in computer simulations are idealizations of the chaos and irreversibility of "nature". Our simulations are classical and nonrelativistic. They have a finite and digital representation. Nevertheless it *is* well-established by now that computational "pseudochaos" provides results which show no important differences from the idealizations of nature in the minds of mathematicians and the real-world observations of experimentalists.

Some of the popular books dealing with chaos and irreversibility seek an understanding of the macroscopic irreversibility of nature in terms of a comprehensive quantum mechanical and cosmological explanation, by linking the present state of the Universe to its "initial conditions". To me it is completely implausible that particular initial conditions, cosmological or not, are at all relevant to understanding the irreversibility present in everyday diffusive, viscous, or conducting flows. None of these ambitious books takes seriously the need for including boundary conditions and constraints in dynamics, which seems to me a crucial ingredient to obtaining irreversible behavior from time-reversible laws. It is clear that computer simulation has been the catalyst for our new understanding of irreversible flows.

There are *many* books addressing irreversibility. Most of them at least mention computer simulation. But there are really only two, Evans and Morriss' *Statistical Mechanics of Nonequilibrium Liquids* and my own *Computational Statistical Mechanics*, which emphasize the primary importance of simulation to a proper understanding of the "reversibility paradox" and the involvement of fractal distributions in resolving it. More mathematical expositions of some of the underlying ideas can be found in Dorfman and Gaspard's books. More philosophical expositions include Coveney and Highfield's, Dudeney's, Hawking's, Penrose's, Price's, and Sklar's. For me, Sklar's is the most interesting of all of them. He emphasizes the classical aspects of the reversibility paradox while simultaneously exploring and expounding a wide variety of alternative points of view.

My own approach is, I think, much the simplest, and proceeds by way of defining nonequilibrium states, in order to reach a compelling understanding of *irreversibility* in terms of the straightforward, but subtle, consequences of *time-reversible* chaotic differential equations. Generating the nonequilibrium states *via* computer simulation is illustrated here in the many example problems. It is my fond hope that the reader will find this approach palatable.

I would like to thank Richard Lim and Robert MacKay for stimulating and encouraging the effort needed to write this book. I also owe a long-standing continuing debt of thanks to my colleagues, friends, and students, for helping me to understand, and to the Livermore Laboratory, the University of California, and universities in Australia, Austria, Germany, Japan, Korea, and Poland, for having provided me with the resources and havens necessary to teaching, research, and good fellowship.

Hideout Canyon Ranch–Red Skunk Ranch, 1998–1999
Stanislaus and Santa Clara Counties, California, USA

Contents

Chapter 1

Time Reversibility, Computer Simulation, Chaos

The moving finger writes, and having writ,
moves on; nor all thy piety nor wit
shall lure it back to cancel half a line;
nor all thy tears wash out a word of it.

Omar Khayyám, as translated by Edward FitzGerald

1.1 Microscopic Time Reversibility and Macroscopic Irreversibility

The "moving finger", the "passing scene", the "unfolding of events". These familiar concepts all convey the notion of motion, linking the present to the past and to the future. Science dissects and idealizes our universal experience of time's passing according to a variety of "physical theories". Any useful physical theory, though approximate and idealized, with limited validity, must be deterministic. A useful theory needs to describe and foretell motion, predicting the unfolding of events with the passage of time.

Fundamental physical theories invariably replace the physical description of nature by a mathematical representation, a set of coupled differential equations. The differential equations give explicit *time* derivatives for all the interesting variables in terms of present values, allowing the "theory" to predict the future or to recover the past by integration. For the usual nonlinear equations the integration is necessarily numerical, and relies on fast computers. In classical physics the variables are macroscopic observ-

1

ables, such as coordinates, velocities, or electric fields. Schrödinger's version of quantum physics provides the time derivative of an intermediate more-abstract "wave function" $\Psi(r,t)$, which can be used to compute observables through a spatial averaging, with probability $|\Psi(r,t)|^2 dr$. Feynman's equivalent formulation, though much closer to classical mechanics, seems to me also to be much more complex. Feynman relates the evolving *quantum* probability to a weighted sum over *all* possible paths $\{r(t)\}$, among which only an infinitesimal fraction could correspond to possible classical evolutions. Because time is a continuous variable, numerical solutions of either the classical or the quantum evolution equations require a sufficiently small timestep Δt. The histories which result are *approximate* descriptions, with the interesting variables given at a discrete number of closely-spaced times, $\{n\Delta t\}$.

Though some philosophers choose to formulate, address, and debate definitions of "time" more detailed than "what a clock measures" we will not. Here I take the concept of time as basic and fundamental. My view of the basic mechanical quantities, time, space, mass, and force, is that these concepts are best grasped intuitively, through examples, and that it is futile and unproductive to seek a deeper philosophical understanding of them. Newton put it thus: "I do not define time, space, place, and motion, since they are well known to all."

Newton's best-known physical theory, classical point mechanics, describes the accelerations ($\{a = \dot{v} = dv/dt = \ddot{r} = d^2r/dt^2\}$) governing the time-evolution of particle coordinates $\{r\}$, as determined by "forces" depending on these coordinates (and sometimes on their velocities $\{v\}$). Notice that this description must implicitly *assume* that directions of increasing spatial coordinates and time are given. *All* physical theories assume, in their formulation, a "subjective" or "psychological" "arrow of time" in order to *define* time derivatives. Because Newton's particles are points, the forces accelerating them represent "action at a distance". Newtonian mechanics is simply a set of differential equations, with variables, constraints, boundaries, and initial conditions which correspond, more or less roughly, to laboratory or astronomical observables.

In parallel to Newton's classical seventeenth-century theory for moving particles, Maxwell's theory for the propagation of electromagnetic waves, and Schrödinger's 1926 theory for the evolution of quantum probabilities, likewise depend upon time through the underlying differential equations which define the corresponding physical theories. And none of these differential equations—Newton's, Maxwell's, or Schrödinger's—displays any explicit distinction between the future and the past. These microscopic equations are all "time-reversible". This means that any coordinate sequence $\{r(n\Delta t)\}$ obtained by solving Newton's motion equations corresponds also to a solution of the *same* equations (but with different initial conditions) in the opposite time direction.

This ubiquitous quality of *time reversibility*, shared by all the fundamental physical theories, is evidently dictated, or at least strongly suggested, to us by nature. This conclusion is based on more than three centuries of careful observation and controlled experimentation. Apart from relativistic corrections, still time-reversible, heavenly bodies move according to Newton's time-symmetric differential equations, $\{m\ddot{r} = \sum F\}$, where the sum of forces includes each mass' gravitational interaction with its fellows. The same motion equations describe simple mechanical experiments on the scale of laboratory experiments.

But the time-symmetric formal reversibility common to both large-scale observation and simple laboratory experiments seems to lack relevance to our intuitive sense of passing time. The apparent incompatibility of "Time's Arrow" with time-reversible physics was nicely captured by a Discover Magazine cover* showing, traffic-sign-fashion, a two-headed arrow bearing the label "One Way". The label expresses the one-way "arrow of time" which sums up our psychological experience of passing time. Our everyday macroscopic experiences are filled with irreversible processes. We see living things age and die; chinaware cracks, chips, and shatters; viscous flows and heat flows generate entropy—they "dissipate" or "degrade" energy into heat. All these dissipative processes can be summarized by phenomenological physical theories included in "irreversible thermodynamics".

Dissipative processes, like aging, fracture, friction, and conduction, are *never* time-reversible, but are nevertheless quite real. In each case the symmetry of time reversibility is broken through interactions which link a system to its surroundings. Though complex, these links can often be modelled by a simple selection of boundary, constraint, and driving forces. Indeed, in computer simulations, just as in nature, no steady dissipative state can be achieved in the absence of such links. They *must* be present, both to govern the flow and to remove the generated heat. Boundary conditions and initial conditions are prerequisites to any solution of dynamical

---

*September 1984. The Alphabetical Bibliography can be found at the end of this book.

equations. Besides transmitting mass, momentum, and energy between a system and its surroundings, these links to the surrounding outside world also transmit *information*—the detailed description of the mass, momentum, and energy transfers. If this transmitted information is discarded, or lost, and degraded into heat, the ability to recapture the past is also lost.

The irreversibility of all macroscopic phenomena leads inexorably to a future where the *order* of the past has been replaced by relative *disorder*. This irreversible growth of disorder can equivalently be described as a loss of information. From this latter point of view it is natural to view macroscopic thermodynamic irreversibility as resulting from a loss of the information required to describe or recover the past. The loss makes it impossible to reverse the flow. Information is a legitimate focus for any physical study, for our information is strictly limited, both by our attention span, and by our capacity for storage. Our best teraflop and terabyte tools for solving the differential equations of physics and storing the solutions are quite limited in their capacities for processing and retaining information. And in a real sense, we cannot hope ever to have more reliable predictions from our physical theories than those produced by computer simulations.

None of the foregoing microscopic physical theories is well-equipped for discussing *irreversible* processes. At first glance it is tempting to declare that no time-reversible theory can ever provide irreversible consequences. Though we will see that this is wrong, the declaration is apparently correct where classical mechanics is concerned, as has been emphasized by a host of scientists. Loschmidt and Zermélo pointed out that the isolated systems of classical mechanics cannot provide a consistent description of irreversible processes. Feynman, Gibbs, Krylov, Ruelle, and Smale are among those who have reïterated this basic point. Gibbs was too cautious to give a precise formulation of the coupling between system and surroundings. But he suggested that the entropy (or available phase-space volume) of a two-part compound system would somehow seek out its maximum value. Gibbs' reticence to discuss detailed mechanisms for maximizing entropy resulted in Krylov's criticism of his contributions. Krylov devoted considerable thought to the "foundations" of statistical mechanics, emphasizing that Liouville's incompressible theorem[†] makes Gibbs' entropy-maximizing arguments invalid, or at least incomplete. Only Krylov's critique of earlier work was published. Although incomplete, and lacking examples, his

---

[†] $\dot{f} = (\partial f/\partial t) + \dot{q}(\partial f/\partial q) + \dot{p}(\partial f/\partial p) \equiv 0$; $f(q, p, t)$ is phase-space probability density.

critical descriptions of Gibbs' work make it clear enough that he had in mind distributions something like the fractal distributions which we now know characterize nonequilibrium states. But he lacked both the computers needed for precise examples and the time to develop and express his own interpretation of irreversibility. From a modern perspective, the ideas which Krylov was able to express appear both incomplete and unconvincing.

Loschmidt pointed out the incompatibility separating time-reversible mechanics from irreversible thermodynamics. He lacked our modern understanding of chaos' rôle in the loss of information responsible for irreversibility. It is a corollary of Liouville's incompressible theorem, that a many-body trajectory confined to a bounded region in phase space will eventually return arbitrarily close to its starting point. Zermélo emphasized that this "Poincaré recurrence" implies that isolated systems cannot display irreversible behavior. Both Loschmidt's and Zermélo's arguments depended upon the concept of precisely-defined trajectories. The existence of chaos shows that this trajectory concept is flawed. In practice, numerical trajectories can neither be computed nor reversed with infinite precision. Additionally, the importance of the boundaries and shielding required to contain an isolated system are generally not discussed. Ruelle and Smale suggested that mechanics needed to be generalized to include boundary effects. A study of boundaries was a natural part of the first realistic nonequilibrium simulations. The important rôle of boundaries is particularly apparent in the example of a confined free expansion, treated in Section 3.10.3.

There is a peculiar view of the importance of *initial conditions* which needs comment. Some philosophers state that Newton's, Maxwell's, and Schrödinger's theories can only be made to provide irreversible solutions by selecting just the right initial conditions. Sklar, for example, repeatedly emphasizes the point of view that choosing the *right* initial conditions is essential to an "understanding" of irreversibility. With time-reversible solutions it is hard to argue for an irreversible future. Any solution which does provide "irreversible behavior" (growing disorder, entropy increase) in the future could be played backward (gaining order, losing entropy). This objection to the simple "explanation" in terms of initial conditions was Loschmidt's. The answer Loschmidt requires became clearer a century later, when chaos theory showed the vital rôle of Lyapunov instability in making the initial conditions irrelevant.

The misguided emphasis on initial conditions likewise ignores the empirical basis of physics. Experiments are useful aids to understanding precisely because microscopic initial conditions are quite irrelevant to the macroscopic outcome. If initial conditions *were* crucial there could be no bodies of knowledge, like physics and chemistry, which describe the evolution of macroscopic behavior independently of the microscopic details.

In addition to this justifiable complaint, choosing the "right" initial conditions appears hopeless whenever, as is usual, the dynamics is "sensitive to initial conditions", meaning Lyapunov unstable. By "Lyapunov instability" I refer to the pervasive growth, exponential in time, of small perturbations of the initial conditions. More details are given in Sections 1.5 and 7.4. In principle Lyapunov instability implies that choosing the "right" initial conditions requires an *infinite* amount of information. This requirement, a consequence of precisely defined trajectories, is *meaningless* from any operational point of view. In practice, double-precision simulations are limited to an uncertainty of about 10^{-28} per coordinate-momentum pair, similar to the uncertainty which governs real experiments when the masses, lengths, and times are respectively of order grams, centimeters, and seconds. This latter observation follows from Heisenberg's "uncertainty principle", and guarantees that any precise effect on the atomic level, due to initial conditions, would disappear in just a few interatomic collision times. A perceptive discussion appears in Ruelle's delightful book *Chance and Chaos*. Double precision is coincidentally also the right choice for continuum simulations, in which microscopic fluctuations of order 10^{-14} are to be ignored.

1.2   Time-Reversible Theories of Irreversible Processes

"Sensitivity to initial conditions" suggests, first of all, that no theory can be judged reasonable if its physical predictions depend sensitively upon the initial data. "Sensitivity" also suggests that a changed less-sensitive theory, perhaps with a more limited goal, might well do better. Such theories can be developed by focussing on another aspect of the solution process, the interaction of the system with its surroundings, as modelled in computer simulations by boundary conditions, constraints, or driving forces. Nonequilibrium systems are typically *open* or *driven* systems, with external sources and sinks of mass, momentum, and energy. Though such systems are more complex than the idealized isolated systems of Newtonian mechanics, the cost of describing this additional complexity is justified by the wide scope of new problem areas such a generalized approach can explore. The simplest useful microscopic models for understanding the everyday world of human experience are time-reversible dynamical theories which include nonequilibrium boundary conditions, constraints, and driving forces.

From both standpoints, computational and analytical, the best physical theories from which to start are (i) the simplest microscopic theory of particle motions, classical mechanics, and (ii) the macroscopic theory of continuum mechanics. Classical mechanics, with boundary conditions, constraints, and driving forces, presently provides a comprehensive description of a wide range of motions, generally thought of as including both the reversible *and* the irreversible types, and ranging in scale from the microscopic level of atomistic mechanics to the macroscopic levels of structural engineering and astronomy. Classical mechanics is a fairly faithful description of physical reality, simpler, and closer to human experience, in most cases, than is its quantum cousin.

A growing appreciation of the complexity inherent in simple theoretical structures has revealed that the task of constructing a comprehensive "unified theory" of "everything" is unattainable. For this reason it is logical to follow Occam's lead, using the philosophical principle of "Occam's Razor" to cut away all but the simplest parts of the candidate theories describing the phenomena of interest. Mechanics, when coupled with boundary conditions, constraints, and driving forces, is enough to explain the symmetry-breaking associated with irreversible processes, and to resolve the conceptual problems associated with conservative mechanics. I like to call the augmented mechanics "thermomechanics" to emphasize its link to thermodynamics and nonequilibrium flows through the explicit incorporation of thermal effects.

Thermomechanics is a direct outgrowth of computation and simulation. When fast computers became generally available, in the 1960s, new problem areas opened up and old analytic approaches could be gracefully abandoned. By the early 1970s, thermomechanics had come into its own as a direct result of computation. Nonequilibrium molecular dynamics was developed in 1972. Nosé's discovery of time-reversible thermostats matching Gibbs' canonical ensemble came in 1984. Much of the subsequent work was devoted to checking that the new methods agreed with Gibbs' statistical equilibrium predictions, as augmented by Green and Kubo's exact formulation of linear transport processes. Direct computer simulation replaced virial series for the pressure and integral equations for the distribution of particle pairs as the simplest path to equilibrium properties. Likewise, computer algorithms largely replaced the construction and analysis of "kinetic equations" for *non*equilibrium problems. The resulting extensions of mechanics to the definition and exploration of *nonequilibrium* systems with special boundary conditions, constraints, and driving forces, would have been incomplete and unrewarding without the computers necessary to solve the underlying differential equations. Let us begin to explore computer simulation by describing the application of fast computers to the task of solving the mechanical motion equations for both microscopic and macroscopic systems.

1.3   Classical Microscopic and Macroscopic Simulation

In classical continuum mechanics, the usual space-and-time-dependent variables are the mass density, velocity, and energy per unit mass,

$$\{\rho(r,t), v(r,t), e(r,t)\}.$$

The motion of a continuum is in principle more complex than that of a system of particles because the dependent variables $\{\rho, v, e\}$ must be known *everywhere*. The time evolution of this set reflects the interdependent flows of mass, momentum, and energy in response to the fields and gradients driving them.

Typical macroscopic computer simulations contain irreversible "constitutive relations". There are two different reasons for this. First, much of the irreversibility we see around us *can* be explicitly and accurately simulated by including Newtonian viscosity and Fourier's heat conductivity. Second, an enhanced *artificial* irreversibility must often be used (artificial viscosities and conductivities are examples) to stabilize numerical techniques. In either case, with "realistic" or "artificial" irreversibility, the simulations are complicated whenever nonlinear effects, leading to chaos, are included. The solutions of the irreversible macroscopic equations can closely resemble the results of laboratory experiments. But, due to their intrinsic irreversibility these macroscopic simulations are often viewed as "less fundamental" than time-reversible microscopic simulations based on particle mechanics. The main criticism levelled at the macroscopic approach *is* its lack of time reversibility. A subsidiary and related aspect of the macroscopic approach is its exclusion of certain fluctuations. The averaging which results in this exclusion has two effects: besides destroying time reversibility it eliminates the complexity associated with extraneous microscopic degrees of freedom. It is only in problems where this complexity is important, like turbulence, that pursuit of the macroscopic approach is bogged down by the complexity characteristic of microscopic representations. The probabilistic nature of quantum mechanics suggests a kind of "averaging" too, but, unlike macroscopic mechanics, Schrödinger's quantum mechanics is completely time-reversible.

Computer simulation solves problems in a way which was novel at the time of the Second World War, and which still meets occasional pockets of resistance. The analytical textbook style of problem solving gives a "solution" described by orthogonal polynomials or series expansions. The computational approach *simulates* the *evolution* of a physical system. The polynomials and expansions are replaced by computer *algorithms*. The computational solution is most likely a time-ordered sequence of coordinate data, supplemented with the evolving values of field variables (stress, heat flux, temperature, and the like). In classical particle mechanics, the trajectories $\{r(t)\}$ describing a solution of Newton's equations $\{F = m\ddot{r}\}$ provide also a reversed, second solution, of the *same* equations, obtained by tracing out exactly the same coordinate values. but in a time-reversed order. In such a time-reversed solution, $\{r(-t)\}$, the particle velocities $\{v \equiv \dot{r} \equiv dr/dt\}$ all change sign, but still obey Newton's equations linking the forces, masses, and accelerations.

How could such a symmetric time-reversible situation reliably describe the irreversible phenomena of the real world? There are several approaches to answering this paradoxical question. But, since the only missing ingredient is the set of initial conditions from which the solution is to be continued, it has been common to "explain" the irreversible behavior by pointing to the special nature of the initial conditions. There is a flaw to this misguided explanation. That flaw is chaos, introduced in the next two Sections and discussed at greater length in Chapter 7.

1.4 Continuity, Information, and Bit Reversibility

Newton's ordinary differential equations of motion describe the motion of mass points subject to forces. The motions which result, $\{r(t)\}$, are typically continuous flows with smooth time derivatives, $\{v(t) = \dot{r}\}$, which obey Newton's differential equations of motion: $\{a = (F/m) = \dot{v}(t) = \ddot{r}\}$. Because the equations are typically nonlinear, and beyond the reach of analytic techniques, a closed-form solution of these equations, giving the particle coordinates as explicit functions of the time, is generally not possible.

As discussed in Section 1.2, the presence of *chaos* in the solution suggests an explanation for the lack of analytic solutions. As time goes by, the "information"—the number of binary bits—required for an accurate analytic solution grows linearly with time. Eventually the required information lies beyond the capabilities of analysis and computation. The gross features of the present and future come to depend upon finer and finer features of the far distant unknowable past. No conceivable improvement of the spatial and temporal resolutions can overcome this problem. The fundamental reason is sobering. It is characteristic of chaotic systems that the most precise experiments or most-carefully-designed simulations cannot probe the future reliably for more than a few collision times. Thus the "determinism" of mechanics is an illusion[†]. It is the *irrelevance* of the initial conditions (due to the randomizing effects of chaos) which makes the systematic study of physics possible. This randomizing—also referred to as "mixing"—corresponds to information loss. As time goes by the link between present and past becomes more tenuous.

A numerical solution of Newton's equations is typically approximate, with limited fourteen-digit precision. The contrast between an ideal continuous solution $\{r(t)\}$, with both the coordinates and the time continuous and precisely known, and a doubly-discretized computer-generated numerical approximation is sharpest if one imagines a solution space in which points are restricted to a regular spatial grid and are evenly spaced in time $\{r(n\Delta t)\}$. In Chapter 2 we discuss Levesque and Verlet's construction of "bit-reversible" doubly-discretized solutions which are *rigorously time-reversible*. Any such numerical trajectory is necessarily periodic, while a continuous trajectory would have to satisfy very special initial conditions in order to achieve periodicity. The *inevitable* periodicity associated with a discrete solution space suggests that such a space cannot be used to describe irreversible flows in isolated systems. Order of magnitude estimates suggest that Poincaré recurrence times are of order $\sqrt{e^{S/k}} = \sqrt{\Omega}$ in a discretized

---

[†]As was well known to Maxwell and Poincaré.

space with Ω discrete states. Such times are effectively infinite (exceeding the Age of the Universe) once the number of particles is of order ten to a hundred. Despite the formal periodicity it is quite possible to describe Lyapunov instability, the sensitivity to small perturbations called "chaos", using the bit-reversible approach.

1.5 Instability and Chaos

Turbulence has long been singled out as a specially "difficult" subject. This characterization of turbulence has arisen from the continuing failure of attempts to predict, or at least to understand, the long-time behavior of complex flows, such as our weather, despite the well-recognized importance of the task. Turbulent instability occurs whenever the decay rate associated with fluid deformations—changes in shape—is sufficiently small. In 1963 Lorenz described his efforts to continue the numerical solution of his now-famous set of three differential equations[§], starting out from intermediate values. He found that this second solution failed to agree with his original one after a fairly short time. Further investigation showed that the mechanism for the disagreement was the *exponentially* unstable loss of information, with the precision required to reproduce a solution of fixed accuracy increasing *exponentially* with the required time, corresponding to the required number of decimal *digits* or binary *bits* increasing *linearly* with time. With Lorenz' work it became "widely known" (to experts) that *most* flows contain this same sensitivity, "Lyapunov instability", to small changes in initial conditions.

Quite typically, both microscopic and macroscopic equations of motion are Lyapunov unstable, meaning that their solutions are very sensitive to small perturbations, so sensitive that such perturbations grow *exponentially* fast, in time. Though it is completely deterministic, and in principle reproducible, the chaos which characterizes Lyapunov instability means that particular precise initial conditions are not a useful concept. This is just as well, inasmuch as the concept of a completely isolated system has no sound basis in physics, where everyday gravitational forces have infinite range. On the other hand, *gross* characteristics of initial conditions *do* describe the spatial variations of macroscopic features, such as the temperature or

[§]$\dot{x} = -\sigma(x - y)$; $\dot{y} = \mathcal{R}x - y - xz$; $\dot{z} = xy - bz$. See Section 4.9.1 for more details.

velocity field. The gross features lead to the reproducible averaged behavior described by the macroscopic physical theories. They also suggest using statistical ensembles to describe the time development of similar systems.

The details of microscopic initial conditions cannot be to blame for irreversible behavior, for the time-reversibility property of microscopic theories is completely independent of these conditions. The only likely possibility for breaking the apparent symmetry of future and past solutions lies in their relative *stabilities*. These stabilities can be analyzed in detail in terms of the Lyapunov spectrum, as is discussed in detail in Chapter 7. The Lyapunov exponents which make up the spectrum describe the *global* time-averaged rates of growth, or decay, of perturbations in the initial conditions. The complete spectrum gives a description of these rates for *all* perturbation directions, not just the most-rapidly-growing one. For Newton's equations of motion both the global *and* the local spectra proceeding forward in time are exactly equal to their time-reversed global or local analogs, regressing *backward*, in time.

Krylov emphasized, that for this reason—the symmetry linking the past and the future—irreversibility cannot be understood from the standpoint of Newton's equations. More recently, Smale and Ruelle have echoed this view, suggesting that the description of irreversible processes requires new generalizations of the classical equations of motion. With the advent of computers making simulations possible, these generalizations were not long in coming. Analyses of the results have revealed an interesting ubiquitous breaking of time symmetry. *Typically*, the *forward* nonequilibrium computer trajectory is *less* sensitive to perturbations—and thus *more* stable—than is the time-reversed backward one. This difference in sensitivity leads both to a symmetry breaking and to a simple geometric understanding of the irreversibility which lies all around us. It also gives rise to singular *fractal* distributions. These are distributions which have no well-defined gradients. Locally smooth in some directions while wildly singular in others, fractals display a pervasive power-law structure on *all* length scales. We will consider the first of many such examples in the next Chapter.

Loschmidt emphasized the paradoxical aspect of time-reversible Newtonian trajectories. And classical mechanics, as originated by Newton, but generalized to include boundaries, constraints, and driving forces, will be the main focus of our interest in reversibility. In relativity theory, electromagnetism, and quantum mechanics, time reversibility is less apparent in the fundamental equations, but is nevertheless present. Schrödinger was intrigued by this problem too. In a 1931 lecture described in his *Science, Theory, and Man* he publicized Exner's sceptical criticism of the view that conventional perfectly-deterministic, and time-reversible, classical mechanics is the only possible model describing "classical" phenomena. So long as energy and momentum are conserved in collisions, a small stochastic contribution to the dynamics *could* also be present, accounting for irreversibility. Exner's explanation, though technically possible, seems implausible, because it fails Occam's test of simplicity. Apart from integer algorithms, like Levesque and Verlet's, finite precision results in computational roundoff error. It seems to me very unlikely that stochastic low-level noise differs in any *significant* respect from this computational error.

1.6 Simple Explanations of Complex Phenomena

Time itself can be viewed as a puzzle, but I choose not to do so. And I also choose to ignore the couplings between mass, space, and time revealed by relativity. For me, Time is a primitive intuitive notion, like Space and Place. I think of time in purely-classical nonrelativistic terms. Time's passage can then be quantified through any periodic motion. It is the result of experience that the exact nature of that motion is immaterial. The difficulties involved in finding a precise and general definition of "time" are not important to an understanding of time reversibility. Simplicity dictates an understanding based on nonrelativistic classical concepts.

At the end of the nineteenth century, and again, toward the middle of the twentieth, some vocal physicists looked forward to finding a unified view of nature. In addition to linking our sensations to the physical world, through understanding consciousness, such a unified view would also require a consistent mathematical description of complex phenomena. The emergence of chaos and complexity renders such a goal obsolete. Gödel showed that most interesting purely-mathematical theories are intrinsically incomplete, unable to decide the truth or untruth of definite statements.

The premature announcements, around 1900 and again around 1950, that "classical mechanics is dead" evidently stemmed from this same obsolete viewpoint of a "complete" theory. If there were some complete and unified view of nature, then more-specialized and restricted special cases of it could perhaps be thought of as second-rate, even if their structure were simpler. Chaos limits the ability of the various theories to overlap.

Physics, chemistry, and biology are intrinsically *different* subjects, rather than special cases of a unified theory. Quantum mechanics is unable to select a particular evolving path. Simulations of *real-world* chaotic processes must *invariably* do just that. Organized biological activity is too complex for a description at the atomistic level. In view of the unattainability of a unified theory, classical mechanics furnishes the best possible basis for understanding problems on its borders with thermodynamics and irreversible fluid and solid mechanics. The exploration and penetration of this artificial perimeter is the main subject of this book.

By now, it is both necessary and commonplace to subdivide knowledge, separating biology from economics and engineering. Within physics classical, quantum, and relativistic mechanics all have their own idealizations, with none of them describing our experience perfectly. I take the point of view that mechanics is an imperfect, but educational model. It is because mechanics' consequences have apt real-life analogs, that this subject is worth knowing. The classical mechanics of isolated systems can be profitably generalized, to describe the interaction of systems with surroundings, nearing the realm of thermodynamics. But the lack of fluctuations in thermodynamics prevents the agreement from being perfect. The only way to distinguish a better theory from its competitors describing the same phenomena, is to wield Occam's Razor, shaving away irrelevant assumptions, and leaving the *simplest* possible explanation as the best. The simplest theories can and do lead to the discovery of complexities as absorbing and interesting as those created by Bach, Brubeck, Mingus, and Monk.

Engineers deal with the application of physical theories to real problems. Their approach is typically totally different to the time-reversible approach of basic physics. Engineering problems include viscosity, heat conductivity, plasticity, and other patently irreversible phenomena based on observation. How does it happen that engineers' alternative approach has lost the fundamental time symmetry of microscopic physics? For one thing, the macroscopic theories used by engineers incorporate averaging, both in space and in time. Their theories are continuum field theories which ignore short-ranged and high-frequency fluctuations. For another, the dependent variables are different too. They reflect the averaging over microscopic details and the process of measurement. Typical variables are temperature, strain rate, and stress, rather than position and momentum, or the wave function. Finally, the systems considered by engineers are seldom isolated. Ordinarily external sources and sinks for heat and work are included. The microscopic analogs of these sources and sinks require generalizing the purely-Newtonian mechanics familiar to physicists.

The familiar irreversible processes which are all around us are not at all similar to the near-equilibrium fluctuations exploited by linear-response theory. Linear-response theory deals with ensemble-averaged infinitesimals. Macroscopic irreversible processes are individual and strongly driven, far from equilibrium, and inherently complex. These far-from-equilibrium conditions require special computer simulation techniques.

1.7 Reversibility Paradox: Irreversibility from Reversible Dynamics

The conflict between basic time-reversible physics and applied irreversible engineering is the "reversibility paradox". For gases, Boltzmann clarified this paradox by showing that averaging, justified by collisional chaos, was an essential part of its resolution. He showed that a statistical averaging of collisions, which ignores any pre-existing correlations and fluctuations, converts the reversible equations governing low-density gas dynamics to the irreversible equations of continuum mechanics. His approximate Boltzmann equation[¶], for the evolution of the single-particle probability density, f_1, makes detailed predictions for the approach to equilibrium, and for the velocity distributions characterizing systems undergoing diffusive, viscous, and conductive dissipation. For dilute gases, the time-development of Boltzmann's approximate single-particle entropy,

$$S_B(t) \equiv -Nk\langle \ln f_1 \rangle \equiv -k\int dr \int dv f_1(r,v,t)\ln f_1(r,v,t),$$

agreed with the predictions of irreversible thermodynamics, opening the way for Gibbs' formulation of statistical mechanics for general systems, but restricted to equilibrium.

Green and Kubo showed that Gibbs' averaging links the irreversible transport coefficients of phenomenological continuum theory to the decay of equilibrium fluctuations. For dilute gases, these results are also equivalent to Boltzmann's. After Green and Kubo discovered linear-response theory, theoretical progress was stalled, awaiting the development of fast computers. The need to understand complex chaotic behavior frustrated

---

[¶] $\dot{f}_1 \equiv (\partial f_1/\partial t)_{\text{collisions}}$. The approximate collision term, $(\partial f_1/\partial t)_c$, is *quadratic* in f_1.

the attempts of analysts. Computers made progress possible again. Recent numerical work has shown that even *few*-body systems show irreversible behavior, on the average, even with rigorously time-reversible motion equations. The irreversibility emerges with great clarity and precision when the small-system results are time averaged.

Computers made it possible to simulate both reversible mechanics and irreversible flows. In the latter case it was necessary to impose boundary conditions or constraints, driving the system from equilibrium. Heat and work had to be incorporated explicitly into the programming. Handily, all this could be done without sacrificing the time reversibility of the underlying equations!

1.8   Example Problems

To illustrate the concepts of time reversibility and chaos we consider here three examples. The first is a two-dimensional area-preserving map. It is a caricature of equilibrium flows obeying Liouville's incompressible theorem. The second is a three-dimensional continuous flow, but with *discontinuous* forces and a simple phase-space structure. The last is a three-dimensional flow with continuous forces and a *complicated* phase-space structure. All three of these problems can exhibit chaotic behavior, with small changes in initial conditions growing exponentially with time. All three are relatively easy to simulate and to visualize. Each is a building block in creating an understanding of the nonequilibrium systems which are emphasized in the following Chapters of the book. The underlying background in mechanics and numerical algorithms, required to generate numerical solutions for the last two problems, is given in Chapter 2.

1.8.1   *Equilibrium Baker Map*

If N coordinates suffice to represent a system's configuration then the configuration at any fixed time is represented by a single point in the corresponding N-dimensional space. The time development of that point defines a one-dimensional "trajectory"—a line—in that same space. Imagine the repeated intersections of such a trajectory with some fixed $(N - 1)$-dimensional surface embedded in the N-dimensional configuration space. The successive intersections with such a surface provide a "Poincaré Section" defining a "mapping" on the surface coordinates, which links each intersection to the next. Such "maps" can be viewed as simplified caricatures of flow problems. Their advantage is a reduction (by one) in the number of coordinates which has to be considered.

Because a *one*-dimensional reversible map can only alternate between two values of the dependent variable, the simplest interesting *reversible* map is necessarily *two*-dimensional. Such a map is analogous to an ordered series of snapshots, each related to the previous by an integration of the motion equations over the time interval Δt by which the snapshots are separated. More complicated maps, with irregular time intervals, correspond to the crossings of particular surfaces in the phase space.

If the underlying motion equations are reversible, the map must likewise be reversible. The two-dimensional Baker Map which we consider here is often used to exhibit chaos, "sensitive dependence on initial conditions", because the separation of two nearby points increases exponentially, in a particular direction, the "unstable" direction. Provided that the initial point is chosen sufficiently randomly, with irrational coordinates for instance, the time-reversible Baker Map shown in Figure 1.1 eventually provides an "ergodic" coverage (coming arbitrarily close to all the points) of the 2×2 square, whether the map goes forward or backward in time. In both directions the map has the same form, but with x and y permuted in the backward map:

$$[-1 < x < +1 \; ; \; 0 < y < +1] \longrightarrow \{x, y\}' = \{(x+1)/2, (2y-1)\};$$

$$[-1 < x < +1 \; ; \; -1 < y < 0] \longrightarrow \{x, y\}' = \{(x-1)/2, (2y+1)\}.$$

This map *reduces* point-to-point separations in the x direction and increases them in the y direction, in both cases by a factor 2. The map is said to be "Lyapunov unstable" because small separation differences in the y direction double at each iteration, corresponding to a maximum Lyapunov exponent of $\ln 2$:

$$\delta y_{n+1} = 2\delta y_n \longrightarrow \lambda_1 \equiv \ln\langle(\delta y_{n+1}/\delta y_n)\rangle = \ln 2.$$

To attain a fixed level of precision in specifying the future iterates of an initial point, additional information would have to be added, in the y direction, and could be discarded, in the x direction, one binary bit per iteration.

Overall, the "equilibrium" Baker Map conserves area, with the stretching (in the y direction) exactly compensating for the shrinking (in the x direction). Evidently the action of the map is "sensitive" to small differences in the stretching direction, so that *not even in principle* could the "true" motion be followed for long. This Baker Map is a rough caricature of motions governed by Lyapunov-unstable Hamiltonian mechanics.

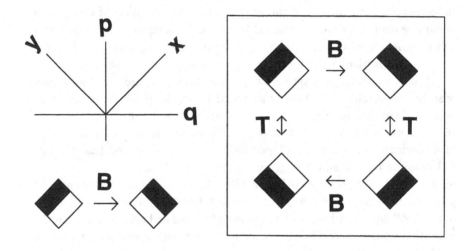

Figure 1.1: The Equilibrium Baker Map, B, in the usual (x, y) coordinates, is shown at the lower left. The mapping B relates the "new" coordinates to the old ones, $B_{xy}(x, y) = (x', y')$. The *rotated* map, $B_{qp}(q, p) = (q', p')$, in $(q, p) = \sqrt{\frac{1}{2}}(x - y, x + y)$ coordinates, is shown at the right. The time-reversal operation T_{qp} shown in the Figure corresponds to changing the sign of the "momentum" p.

When the exponential instability associated with stretching is represented by a limited-precision computer, it can give rise to a fixed y coordinate, at $y = \pm 1$, and a fixed x coordinate, $x = \pm 1$, so that the resulting "fixed point" repeats forever. Likewise, there are particular initial conditions which give rise to (unstable) periodic cycles of two, three, four, ... iterations. For example, the twice-iterated Baker Map has a cycle connecting the two (x, y) points $(-1/3, +1/3)$ and $(+1/3, -1/3)$. On the other hand, adding a small amount of random noise, in the eighth, tenth, or twelfth significant figure, is enough to destroy these artificial cycles.

This equilibrium Baker Map, though simple, already illustrates a disquieting feature, due to its singular nature along the line $y = 0$: any *rational* value of y, of the form $\pm n/2^m$, is eventually mapped to the singular line. For any fixed m these solutions, like those of the periodic cycles, have measure zero in the two-dimensional space. General *irrational* points are instead mapped in an "ergodic" manner, eventually coming arbitrarily close to any (x, y) point in the square. Despite this uniform coverage of the square, the Baker Map is a very poor random number generator. In both the (x, y) and the rotated (q, p) representations it is easy to show that the products of successive iterates are strongly correlated:

$$\langle xx' \rangle = 1/6 \; ; \; \langle yy' \rangle = 2/3 \; ; \; \langle qq' \rangle = \langle pp' \rangle = 5/12.$$

All these correlations would vanish if the (x, y) points or the rotated (q, p) points could truly be chosen *randomly*.

The odd sensitivity to infinitesimal insignificant "information" which has no significance, being infinitesimal in scale, can plainly have nothing to do with physics. The sensitivity is a consequence of the vagaries of the real number system, in which any two rational numbers, no matter how close together, are separated by an infinite continuum of irrationals as well as a set with the lesser cardinality, $\aleph_0$ in Cantor's theory of infinite sets, of rational numbers. Because the Baker Map is *so* simple some computer programs, on some computers, will not generate sufficient internal noise to avoid settling onto one of the Map's fixed points at $(x, y) = \pm(1, 1)$. In an accurate calculation this could not happen because these fixed points are unstable, with displacements in the y direction doubling at each iteration.

The disparity in evolution, between the rational periodic orbits and general irrational initial points, seems less serious if we consider exactly the same map, expressed now in terms of the horizontal and vertical coordinates, $\{q, p\}$. The motivation for this rotation is the physicist's notion of time reversibility: coordinate values $\{q\}$ should be traced out backward in time simply by (i) reversing the signs of the corresponding momenta $\{p\}$ and (ii) choosing the right initial conditions. The rotated (q, p) coordinates are related to the original (x, y) coordinates by a 45° clockwise rotation:

$$q \equiv (x - y)/\sqrt{2} \; ; \; p \equiv (x + y)/\sqrt{2}.$$

Thus the periodic cycle of the twice-iterated map links the two (q, p) points $(\pm\sqrt{2}/3, 0)$. In the new (q, p) coordinate system the map satisfies the usual physicists' expectation for a time-reversible map:

$$B_{qp}(q, p) = (q', p') \longrightarrow B_{qp}(q', -p') = (q, -p).$$

Explicitly, this rotated Baker's Map has the piecewise-linear, but *irrational* form:

$$q < p \longrightarrow \{q, p\}' = \{+(5q/4) - (3p/4) + \sqrt{9/8}, -(3q/4) + (5p/4) - \sqrt{1/8}\};$$

$$p < q \longrightarrow \{q, p\}' = \{+(5q/4) - (3p/4) - \sqrt{9/8}, -(3q/4) + (5p/4) + \sqrt{1/8}\}.$$

See Figure 1.2 for a sequence of points generated using this map. We will come back to simple mappings again, in discussing dissipative analogs of the present area-preserving conservative Baker Map. Though simple, these maps have considerable pedagogical value.

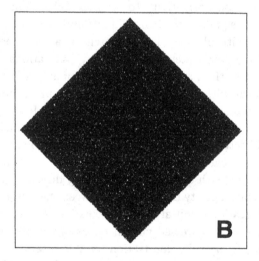

Figure 1.2: 100,000 points generated by the (q, p) Baker Map beginning with initial point at (0.50,0.00).

Evidently a probability density $f(x, y)$ which is constant is preserved exactly by the Baker Map. Each area element $dxdy$ is mapped to a new element with exactly the same area, $dx'dy' = (dx/2)2dy$. Thus the constant-density solution $f(-1 < x < +1, -1 < y < +1) \equiv 1/4$ is *stationary*. Apart from the lines with special rational y coordinates, $y = n/2^m$, along which trajectories approach the fixed points $(-1, -1)$ and $(+1, +1)$, and the various unstable periodic cycles, a general irrational point in the 2×2 square eventually converges, in a "coarse-grained" sense, to the constant-density solution. This means that the integrated density inside *any* square box of sidelength ϵ (that is, the probability of occupying that box) approaches the limiting value $\epsilon^2/4$.

Because in principle the mapping process does not destroy any information, and can certainly be inverted—even precisely reversed, in the (q, p) representation—the solution becomes constant only in the limit of an infinite number of iterations. In practice, with a fixed number of significant figures, $\log(1/\epsilon)$, the mapping eventually produces a probability density which is essentially constant in cells of width and height exceeding ϵ. Increasing the precision of the representation and the mapping reduces the discrepancy between averages computed with the constant density $f = 1/4$ and the coarse-grained approximation. We will see, in the next Chapter, that a dissipative version of the Baker Map, a caricature of *non*equilibrium systems, provides an infinitely-detailed "fractal" solution, rather than the constant density f found here. Fractals display structure on *all* scales, no matter how fine. Because this behavior is not only unpredictable in principle, but is also too detailed for accurate description, averaging, in both space and time, is inevitably required. This averaging leads to the "coarse-grained" representations envisioned by Gibbs.

1.8.2 *Equilibrium Galton Board*

A more realistic model of an irreversible process has continuous "chaotic" trajectories, so that numerical solutions are "Lyapunov unstable". The "Galton Board" is such a model. It is the *simplest* analog for field-driven diffusion. The Galton Board[||] is named for Sir Francis Galton, who used a similar laboratory device, beginning in 1873. He demonstrated the familiar binomial distribution, which results when falling particles are scattered to the right or left through a "Board" of scatterers. The probability distribution which results is the familiar binomial distribution:

$$\mathrm{p}(n_r, n_l) = [(n_r + n_l)!/(n_r!n_l!)](1/2)^{n_r+n_l}.$$

In both Galton's laboratory demonstrator and the corresponding com-

---

[||]Sometimes called the periodic "Lorentz Gas".

putational model, a single particle moves through an array of hard-disk scatterers. In computation it is convenient to treat the moving particle as a mass point, scattering from a lattice of fixed particles of *radius σ*. Equivalently, it is possible to treat this problem as the spatially-periodic dynamics of *two* moving particles, each of *diameter σ*, with one particle accelerated to the right and the other to the left. Figure 1.3 illustrates the first of these arrangements, a moving point, with the scatterers arranged in a regular "triangular" lattice.

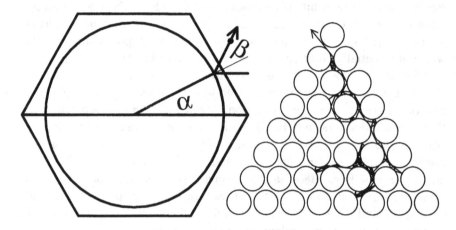

Figure 1.3: Portion of a Triangular Lattice with an *equilibrium* trajectory— $g \equiv 0$, where g is the accelerating field strength. Each of the 157 collisions shown defines two angles, (α, β), giving its location and the post-collision direction of motion, as indicated in the hexagonal "unit cell" at the left.

It is convenient to describe the collisions with two angles: α gives the location of a collision relative to the field direction; β gives the direction of the scattered trajectory, relative to a vector *from* the fixed scatterer *to* the moving point, just *after* each collision. The sequence of collisions with the curved surfaces of the scatterers provides chaos. This chaotic instability means that small perturbations in the initial coordinates and momenta of the moving particle, $\{\delta q, \delta p\}$, will grow exponentially in the time, as $e^{\lambda t}$. Just as in the Baker Map, there is both a growing "unstable" direction and a shrinking "stable" one, so that a continuous flow of new information (in the growing direction) is required to operate the map. This quite typical "sensitive dependence on initial conditions" causes the particle to experience, as time goes on, all possible collision types: $\{0 < \alpha < \pi \; ; \; -1 < \sin\beta < +1\}$. If all of space is represented by a single unit cell, the limiting spatial distribution within that cell becomes completely uniform in the absence of any accelerating field. These rather obvious features were proved by Sinai, and can be convincingly demonstrated by a short computer program simulating the motion.

The simplest computer program simulating the time development of the Galton Board advances the moving particle step-by-step along the field-free straight line trajectory,

$$x_+ = x_0 + \dot{x}_0 \Delta t \; ; \; y_+ = y_0 + \dot{y}_0 \Delta t \; ; \; \dot{x}_+ = \dot{x}_0 \; ; \; \dot{y}_+ = \dot{y}_0,$$

or the field-induced parabola,

$$x_+ = x_0 + \dot{x}_0 \Delta t + (g/2)\Delta t^2 \; ; \; y_+ = y_0 + \dot{y}_0 \Delta t \; ; \; \dot{x}_+ = \dot{x}_0 + g\Delta t \; ; \; \dot{y}_+ = \dot{y}_0,$$

checking to see whether or not a collision, with either a periodic boundary or the scatterer boundary, has occurred. In the event of a "collision" with a periodic boundary, the particle is simply replaced, in the cell, with a new velocity, $(\pm\dot{x}, \pm\dot{y}) \rightarrow (\pm\dot{x}, \mp\dot{y})$, and the trajectory is continued. In the event of a scatterer collision, the particle is returned to its location at the previous timestep, and the radial momentum component (in the frame of the scatterer) is reversed. Figures 1.4 and 1.5 show the time development of two problems: (i) the uniform field-free distribution among collision types, which also corresponds to a uniform distribution in space; (ii) a *non*uniform distribution of collision types induced by an accelerating field g. In both cases I use $\sin\beta$ rather than β to catalog the calculated collision sequences. This gives a *uniform* distribution in the field-free case because the *relative* collision rate is proportional to $\cos\beta d\beta = d(\sin\beta)$. The steady-state motion characteristic of a *thermostated* Galton Board is simulated in Section 5.9.1.

For any fixed coarse-graining of the $(\alpha, \sin\beta)$ space the deviations from the field-free uniform distribution eventually depend inversely on the square root of the sampling time, as expected for a random process. From the theoretical standpoint, the entire distribution could be generated by an alternative, much more cumbersome, approach. We could solve Liouville's *partial differential equation* for the time-development of the probability density[**]. For the present model it is quite clear that the trajectory dynamics provides a much simpler route to the steady-state distribution.

[**] $\dot{f} = (\partial f/\partial t) + \dot{q}(\partial f/\partial q) + \dot{p}(\partial f/\partial p) \equiv 0 \; ; \; f \equiv f_N(q, p, t) \; ;$ for details see Section 2.5.

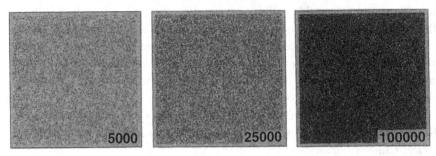

Figure 1.4: $(\alpha, \sin\beta)$ pairs for 5000, 25,000, and 100,000 successive collisions in the field-free case, with $g = 0$. The scatterer density is 4/5 the
close-packed. The moving particle has unit mass and kinetic energy 0.50.

In the absence of "thermostat" or "control" forces, discussed in Chapter
2, the field-driven distribution never converges. The field-driven Newtonian
motion of the Galton Board has no stationary state. The collision pairs
$(\alpha, \sin\beta)$ shown in Figure 1.5 were generated by advancing coordinates
chosen randomly within the unit cell, with initial velocity zero, and plotting
the resulting scattering collisions whenever the kinetic energy was within
the limits $0.495 < K < 0.505$. The resulting distribution, though generated
with classical Newtonian mechanics, shows a strong family resemblance
to the "fractal" distributions characteristic of nonequilibrium steady-state
systems discussed in this book.

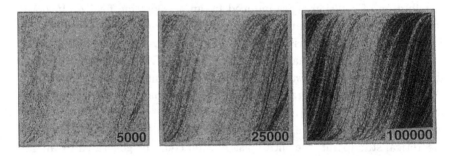

Figure 1.5: $(\alpha, \sin\beta)$ pairs for 5000, 25,000, and 100,000 collisions generated
from randomly chosen zero-velocity initial conditions but with a nonzero
field, $g = 0.5$. The scatterer density is 4/5 the close-packed density. The
scatterer diameter is unity and the moving particle has unit mass. Those
collisions are shown which had kinetic energies between 0.495 and 0.505.

How could one make an *integer* model of the Galton Board, with a discrete coordinate space? The periodicity associated with a discrete coordinate space seems fundamentally opposed to chaos, but, as we shall see, Levesque and Verlet were able to construct a time-reversible algorithm for chaotic problems by using an integer coordinate space and a continuous potential. How could one construct a discrete integer-space version of the Galton Board? Evidently this *could* be done, by using a steepened version of the potential—like that used in Section 2.11.2, rather than the discontinuous hard-disk scattering law considered here.

1.8.3 *Equilibrium Hookean Pendulum*

The simplest chaotic problems, suitable for computer simulation, and free of the singularities of the Baker Map and Galton Board, involve pendula. A double pendulum is relatively easy to build[tt]. Its analytic counterpart is a familiar chaotic problem free of any singularities. A conceptually simpler model, which is advantageous for studying the dependence of the local (instantaneous) Lyapunov exponents on the coordinate system, is a "Hookean pendulum" with its length governed by a Hooke's-Law spring. The Hamiltonian governing the motion can be written in either Cartesian coordinates $\{x, y\}$ or plane polar coordinates $\{r, \theta\}$:

$$\mathcal{H}_C = [p_x^2 + p_y^2]/2m + mgy + (\kappa/2)(\sqrt{x^2 + y^2} - d)^2.$$

$$\mathcal{H}_P = [p_r^2 + (p_\theta/r)^2]/2m + mgr\sin\theta + (\kappa/2)(r - d)^2.$$

As is explained more fully in the next Chapter, the equations of motion, in either coordinate system, follow from the Hamiltonian flow equations:

$$\{\dot{q} = +(\partial\mathcal{H}/\partial p)_q \; ; \; \dot{p} = -(\partial\mathcal{H}/\partial q)_p\},$$

where q is a coordinate in the Hamiltonian (x or y or θ or r), and p is the corresponding momentum (p_x or p_y or p_θ or p_r).

For convenience, choose the pendulum mass m, the rest length d, and the gravitational field strength g equal to unity with the force constant κ equal to four. This last choice promotes the Lyapunov-unstable coupling between

[tt]Ted Hillyer built one for me, based on James Yorke's lecture-demonstration model.

the vertical and horizontal oscillations of the pendulum. The equations of motion follow by differentiation:

$$\{\dot{x} = p_x \; ; \; \dot{y} = p_y \; ; \; \dot{p}_x = -4(x/r)(r-1) \; ; \; \dot{p}_y = -1 - 4(y/r)(r-1)\};$$

$$\{\dot{r} = p_r \; ; \; \dot{\theta} = p_\theta/r^2 \; ; \; \dot{p}_r = (p_\theta^2/r^3) - \sin\theta - 4(r-1) \; ; \; \dot{p}_\theta = -r\cos\theta\}.$$

The two sets of motion equations provide alternative descriptions of the *same* motion, so that a comparison furnishes a good check of the numerical integrator. The Runge-Kutta integrator described in Section 5.2 is the simplest useful choice. Both energy conservation:

$$\dot{\mathcal{H}} \equiv \sum[\dot{q}(\partial\mathcal{H}/\partial q) + \dot{p}(\partial\mathcal{H}/\partial p)] \equiv 0,$$

and time reversibility—(i) running forward for a time t, (ii) reversing the momenta, (iii) advancing the dynamics (and so going "backward") for time t, and (iv) reversing the momenta once again, finally arriving at the initial configuration with the initial momenta:

$$\{+q, +p\}_0 \xrightarrow{\text{(i)}} \{+q', +p'\}_t \xrightarrow{\text{(ii)}} \{+q', -p'\}_t \xrightarrow{\text{(iii)}} \{+q, -p\}_0 \xrightarrow{\text{(iv)}} \{+q, +p\}_0,$$

are additional useful checks of the numerical work.

Any solution of the pendulum problem can be visualized as a trajectory in the four-dimensional space, "phase space", with axes $\{q_1, q_2, p_1, p_2\}$, where a single "point" specifies the entire state of the system. For certain initial conditions the Hookean pendulum's motion is chaotic. If the evolution of a second pendulum problem is followed in the same phase space, with nearly the same chaotic initial condition, the magnitude of the "offset vector" connecting the two nearby trajectories,

$$\delta_C \equiv \sqrt{\delta x^2 + \delta y^2 + \delta p_x^2 + \delta p_y^2},$$

in the Cartesian case, and

$$\delta_P \equiv \sqrt{\delta r^2 + \delta\theta^2 + \delta p_r^2 + \delta p_\theta^2},$$

in the polar case, will then eventually grow as $e^{\lambda t}$. The long-time average of this local growth rate λ defines the "largest Lyapunov exponent", $\lambda_1 \equiv \langle\lambda\rangle$. It characterizes the chaotic motion. Although this growth rate must be independent of the choice of coordinates, when averaged over a sufficiently long time interval, the instantaneous values of the growth rate *do* depend upon the chosen coordinate system, so that the local Lyapunov exponents are properties of the description rather than the physical system. Figure 1.6 shows probability densities for "local" (time-dependent) values of the Lyapunov exponents $\{-5 < \lambda < +5\}$, using both the cartesian and the polar-coordinate systems. Computation of the complete spectra is discussed in Section 7.4.

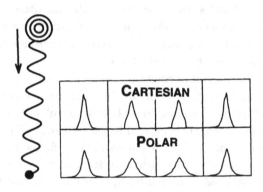

Figure 1.6: Probability distributions for the four Lyapunov exponents of the Hooke's-Law pendulum shown at the left. The corresponding Hamiltonian has the value, $\mathcal{H} = 1$, with the initial conditions $(x, y) = (0.00001, 1.0)$; $(p_x, p_y) = (0.0, 0.0)$. The *mean* values of λ_2 and λ_3 vanish. By symmetry $p(\pm\lambda_1) = p(\mp\lambda_4)$ and $p(\pm\lambda_2) = p(\mp\lambda_3)$.

1.9 Summary

The time reversibility of fundamental physical theories might seem to preclude the irreversibility which is characteristic of physical reality. But the presence of chaos leads to two important effects, which together negate that idea. First, the chaotic Lyapunov instability inherent in single-trajectory solutions rules out any useful link between a long-time trajectory and the initial conditions. Boltzmann, Gibbs, and Maxwell all emphasized the importance of a probabilistic interpretation of dynamics reflecting this underlying instability. The *positive* Lyapunov exponents describe this instability, which corresponds to the continual creation of new information. Second, existing information is typically destroyed in chaotic evolutions. The *negative* Lyapunov exponents describe this stabilizing feature. Though both positive and negative Lyapunov exponents coexist in chaotic problems the negative exponents have the upper hand. They imply a stable convergence as a result of the overall destruction of system information.

Over a long time, the equilibrium situation corresponds to an exact balance of these tendencies, while the nonequilibrium situation invariably corresponds to an enhanced stability and to a net destruction of information, equivalent to a net increase of the external surroundings' entropy. These ideas will be illustrated in detail in Chapter 5.

The irrelevance of the initial conditions, due to chaos, certainly suggests that an alternative to the study of detailed trajectories could be profitably taken up. Following the successful ideas Gibbs formulated at equilibrium, detailed in Chapter 3, perhaps we could follow nonequilibrium probability densities directly in time, making it unnecessary to follow detailed trajectories? If such an ensemble approach were successful, it would provide a description of an averaged approach to equilibrium, followed by Gibbsian fluctuations. Numerical work indicates instead that simple trajectory analyses provide the simplest route by far to nonequilibrium properties, despite the evident approximate nature of their resemblance to "real trajectories".

Real trajectories, as Joseph Ford was fond of pointing out[‡‡], are figments of the imagination, once chaos is considered. Despite the promise of chaos' help in understanding the irrelevance of initial conditions, the impossibility of constructing a chaotic trajectory is a real fly in the ointment, imposed upon us by the real number system. Cantor's set theory established that the rational numbers which we use in computation are a negligibly-small, but uniquely important, subset. In a *mathematical* sense there is a qualitative distinction between systems with finite and infinite numbers of states. In a *practical* sense this distinction appears to be insignificant. Although it appears that this distinction has no practical importance, it is precisely as unsettling as the question of free will when we wish to relate computation to the real world around us. To date Gödel has come the closest to explaining difficulties of this kind.

[‡‡]See his Chapter in *The New Physics*, published by Cambridge University Press (1989).

Chapter 2

Time-Reversibility in Physics and Computation

To see a world in a grain of sand, ...
and eternity in an hour.

Wm. Blake

2.1 Introduction

All of the fundamental differential equations of mathematical physics—
Einstein's, Hamilton's, Lagrange's, Maxwell's, Newton's, Schrödinger's—
are "time reversible". The dependent variables can be the particle or continuum coordinates of classical mechanics, or the electromagnetic field, or
the wave-function underlying the probabilities of quantum mechanics. The
reversibility of all these theories differs qualitatively from the *irreversibility*
of thermodynamics. In this book I consider "time reversibility" in detail,
for systems based on classical mechanics and thermomechanics. What does
time reversibility mean for these systems? Simply that all possible "solutions" of the fundamental equations—time histories of particle or field
variables—can be followed either forward or backward in time *without any
change in the equations themselves*. Reversing time requires only a change
in the initial conditions.

By contrast, precisely reversible *computation* is quite rare. Computational roundoff errors accumulate. Typically there is no simple relation
linking the errors in a reversed trajectory to those of the forward trajectory. The exponential growth of these differences frustrates attempts to

reverse trajectories for more than a few collision times. In parallel computation, where a problem is divided up among several processors, the results can lack reproducibility for a different reason. The dependence of sums on the order of their evaluation (with $a + b + c$ differing from $c + b + a$, for instance) can thwart reproducibility. Similar disparities occur between different compilers and different hardware, even when only a single processor is used. I know of only one way to avoid the irreversibility inherent in this general sensitivity to roundoff error. This is to use a scheme like Verlet and Levesque's "bit-reversible" "leapfrog" dynamics, in which *integer arithmetic* coupled with a time-symmetric algorithm gives trajectories rigorously reversible, "to the very last bit". It appears that this approach is both stable and accurate for the equilibrium flows described by Liouville's incompressible theorem. The bit-reversible algorithm is fully described in Section 2.3. The flows it generates are equilibrium isoenergetic flows, with no natural tendency toward the dissipative contraction which characterizes nonequilibrium systems.

We will see that in only slightly more general situations, both at and away from equilibrium, phase volume is *not* conserved, even in principle. Stationary nonequilibrium states are quite typically characterized by *many-to-one* attractive mappings in phase space. Evidently no such contracting flow could be described by a bit-reversible algorithm. Thus the motion equations of the *non*equilibrium systems which specially interest us in this book *cannot* be solved by bit-reversible methods.

Computer simulations can extract detailed "experimental data" from physical theories in a way which physical experiments cannot. But in order to relate the results of these simulations to the world around us it is necessary that the simulation variables have real-world physical analogs. So far we have discussed mechanics. Forces and stresses in mechanical simulations correspond quite naturally to their experimental analogs. In broadening mechanics to include thermodynamics *temperature* is the essential new concept. In Section 2.8 we will discuss temperature, and its introduction into mechanics, through thermostats, from the perspective of time reversibility. Though there is nothing specially peculiar about heat transfer or thermostated trajectories, the nonequilibrium probability distributions to which the trajectories correspond *are* profoundly different to their equilibrium counterparts. As a consequence of an overall contractive character, these nonequilibrium distributions become *fractal attractors*. Their fractal nature means that they have structure on all length scales.

2.2 Time Reversibility

Though time reversibility would seem to be a familiar intuitive notion, the details can be confusing, and the definitions can seem all too vague, when it comes to applications. In part, this stems from the fact that dynamics is described by time-reversible *differential* equations and is imagined as the *continuous* time development of a set of *precisely-defined* coordinates. Because analytic approaches fail for chaotic problems, practical solutions are necessarily numerical and discontinuous, with a typical precision of fourteen decimal digits. Real continuity, in both space and time, and with infinite precision, is never present in such a "solution" of the equations because numerical solutions are necessarily finitely-expressible, or "coarse-grained". Such finite approximations nearly always lack the time reversibility of the underlying differential equations.

The usual numerical solutions, representing attempts to describe functions of a continuous time variable, inevitably involve some approximation or truncation. Series of coordinate sets, given at discrete times, truncated Taylor or Fourier series in the time, and movies of the motion, are example representations. It appears that the time reversibility of a particular approximate numerical solution depends not only on the form of the underlying differential equations, but also on the way in which the solution is approximated and presented. The picture becomes cloudier still when the evolution of not just a trajectory, but rather a fractal-density *ensemble* of systems, is described.

Let us focus on the simpler case, not an ensemble, but rather the evolution of a single dynamical system. Consider the following representative forms which dynamical "solutions" of the system trajectory can take:

    (i) A set of movie frames;

    (ii) An ordered set of integer coordinates;

    (iii) An ordered set of floating point coordinates;

    (iiii) A truncated Fourier time series;

    (v) A truncated Taylor time series.

Evidently any movie is "time-reversible" in a simple nontechnical sense, whether or not the underlying equations or map (if any) are reversible, because the movie frames could equally well be projected in reverse order. For differential equations this type of reversal is equivalent to changing the sign of the timestep $(+\Delta t \rightarrow -\Delta t)$ in using a computer algorithm to approximate the solution of a differential equation. This type of reversal has been called "invertibility" and "reversibility with time inversion" by Illner and Neunzert. Such reversals are not at all what physicists mean by "time reversibility". For a physicist, the fundamental differential equations describing the forward evolution in time are necessarily *identical* to those going backward in time. Only the initial conditions, *not* the equations, can differ in the "reversed" flow. To clarify this observation consider the approximate harmonic-oscillator trajectory shown in Figure 2.1 (Segment #1, at the *top right* of the Figure).

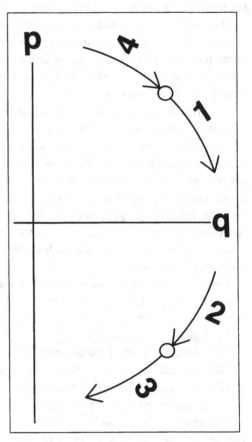

Figure 2.1: Harmonic oscillator trajectory segments. The time-reversal of any trajectory segment is a two-step process: (i) Reflect the segment $\{(+q, +p)\} \rightarrow \{(+q, -p)\}$; (ii) Reverse the time-ordering.

The oscillator coordinate q is advanced in time from its initial value q_0 and the momentum p, initially p_0, equal to $m\dot{q}$, changes accordingly. Newton's formulation of the oscillator motion can be written as a single second-order ordinary differential equation for q, or as an equivalent set of two coupled first-order equations for (q, p):

$$m\ddot{q} = -\kappa q \longleftrightarrow \{\dot{q} = v = p/m \; ; \; m\dot{v} = \dot{p} = -\kappa q\}.$$

The time-reversed trajectory (Segment #2 at the *bottom right* of the Figure) goes through *exactly the same* coordinate values, but in reversed time order, and with momenta which are changed in sign at each coordinate value,

$$+p_{\text{forward}}(q) = -p_{\text{reversed}}(q) \longleftrightarrow +v_{\text{forward}}(q) = -v_{\text{reversed}}(q).$$

To recapture the "past" history of the oscillator, Segment #4, requires a two-step process. First, integrate forward from the time-reversed initial point $(+q_0, -p_0)$, obtaining Segment #3. Reverse both the momenta *and* the time-ordering of these points to obtain the desired "past", Segment #4.

2.3   Levesque and Verlet's Bit-Reversible Algorithm

Let us choose a simple example problem governed by Newtonian mechanics, with $\{a = (F/m)\}$. Discretize the time evolution of the chosen Newtonian system, and represent its spatial coordinates as *integers*, given at equally spaced times. Levesque and Verlet pointed out[*] that such sets of integer coordinates, describing the "time" development of a dynamical system, can be generated with a special "bit-reversible" algorithm, such as the simple Størmer-Verlet centered-difference scheme, but with *integer*-valued coordinates:

$$\{a_0(\Delta t)^2 \equiv q_+ - 2q_0 + q_- = [(F/m)_0(\Delta t)^2]_{\text{integer}}\}.$$

Here the subscripts indicate three contiguous times, separated by intervals Δt. *The righthand side is to be truncated to an integer.* This algorithm is exactly "bit-reversible". This means that the full—but finite—precision of the coordinate data is regained in the reversed trajectory. The price for this exact coordinate reversibility is *approximate forces. All* the combinations $\{(F/m)_0(\Delta t)^2\}$ are rounded off to integer values. With this approximation for the forces, the reversal of the numerical coordinate sets is *exact*, to the very last bit. Levesque and Verlet's symmetric algorithm can evidently be extended, equally well, either forward or backward in time, starting

---

[*]in the Journal of Statistical Physics **72**, 1993; see also Kum and Hoover, in **76**, 1994.

with any two contiguous coordinate sets. The algorithm is patently time-reversible. The numerical values which it generates are replayed *exactly*, rather than approximately, in any time-reversed trajectory.

As an example, consider again the one-dimensional harmonic oscillator, with the combination $(\kappa/m)_0(\Delta t)^2$ arbitrarily set equal to unity. A sample bit-reversible solution of the centered-difference scheme is the repeating sequence of integer coordinates:

$$\{q\} = \{-2, -1, +1, +2, +1, -1, -2, -1, +1, +2, +1, -1\}.$$

Very few sets of dynamical equations are precisely reversible, "bit-reversible" as algorithms. But there *are* a few other examples:

$$\{q_{++} - q_+ - q_- + q_{--} \equiv [(\Delta t/2)^2(5a_+ + 2a_0 + 5a_-)]_{\text{integer}}\}.$$

It is much more common to find time-reversible algorithms which are *not* bit-reversible. Consider the set of approximate first-order implicit harmonic-oscillator equations:

$$\{q_+ = q_0 + (\Delta t/2)(v_+ + v_0) \; ; \; v_+ = v_0 - (\Delta t/2)(q_+ + q_0)\}.$$

This approximation to the differential system $\{\dot{q} = +v \; ; \; \dot{v} = -q\}$, though patently *time*-reversible, is not *bit*-reversible. This is apparent from the analytic solution of the implicit difference equations:

$$q(n\Delta t) \propto e^{in\alpha} \; ; \; \cos(\alpha) = [1 - (\Delta t/2)^2]/[1 + (\Delta t/2)^2].$$

A sample solution of the implicit oscillator equations, for the special choice $(\Delta t)^2 = 4/3$, is the repeating sequence of coordinate-velocity pairs (q, v):

$$\{(q, v)\} = \{(-2, 0), (-1, +\sqrt{3}), (+1, +\sqrt{3}), (+2, 0), (+1, -\sqrt{3}), (-1, -\sqrt{3})\}.$$

With stationary boundary conditions any bounded bit-reversible solution must also be, in principle, periodic. Eventually, because the number of integer-valued state points is finite, the initial state *must* repeat exactly, so that the "dynamics" consists of a single periodic orbit. Figure 2.2 shows the time evolution of a 36-particle system generated in this way. A reversal of all the momenta after another 100,000 time steps—interchanging $\{q_-\}$ and $\{q_+\}$ while leaving $\{q_0\}$ unchanged—leads, precisely and exactly, to the time-reversed initial condition 100,000 time steps later. See the Figure.

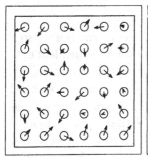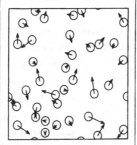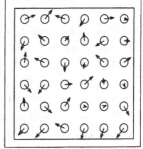

Figure 2.2: Initial, final, and time-reversed 36-particle bit-reversible configurations. The initial configuration is the regular structure shown at the left. The central "final" configuration occurs 100,000 time steps later. That same configuration, but with the velocities reversed, gives the rightmost configuration after another 100,000 iterations. Note that the initial configuration is recovered exactly, but with all the velocities changed in sign from the initial values. See Kum and Hoover (1994).

Suppose that the floating-point version of the leapfrog algorithm, rather than the bit-reversible integer algorithm, had been applied to this same many-body problem. Then Lyapunov instability would immediately have destroyed the exact step-by-step reversibility of the numerical trajectory. This destruction of information would have been caused by the *exponential* amplification of the inevitable roundoff errors. Whether or not the distinction between *exactly periodic* bit-reversible orbits and their irreversible relatives has any practical consequences is not known. The overall usefulness of computer simulations, *exactly* reversible or not, argues strongly that it does not.

Although in practice simple stepwise algorithms are usual, in principle there are many alternative representations of dynamical trajectories. Truncated Fourier series suggest the absence of Lyapunov instability (which is actually present in real problems) because such series imply periodicity. Truncated Taylor's series are also limited in usefulness, by their finite convergence radii. High-order series solutions are not very useful when discontinuities are present. Hard-sphere collisions are discontinuous, for instance. *Any* useful forcelaw *necessarily* has singular high-order derivatives in order to keep the range of the interaction finite.

2.4   Lagrangian and Hamiltonian Mechanics

Lagrange's and Hamilton's generalizations of Newtonian mechanics are specially useful for describing constrained systems. Coordinate-dependent constraints can be incorporated in a governing Lagrangian $\mathcal{L}$ automatically. Thus, *constrained* systems (such as those composed of molecules whose fixed shapes define a rigid structure) are best described by *Lagrangian* mechanics. Systems for which energy is more fundamental than force (quantum mechanics is the best example) are best treated by *Hamiltonian* mechanics. For the *nonequilibrium* thermomechanical systems either of these classical forms of conservative mechanics can furnish a useful beginning.

Let us begin with Lagrangian mechanics. Lagrangian mechanics replaces the Cartesian $\{x, y, z\}$ coordinates of Newtonian mechanics by "generalized coordinates", $\{q\}$. Generalized coordinates can, for example, include bond lengths, angles, or normal-mode vibrational amplitudes. Lengths and angles are particularly useful in formulating geometric constraints. Usually the Lagrangian $\mathcal{L}$ governing the motion is the difference between the kinetic and potential energies, $\mathcal{L}(q, \dot{q}) = K - \Phi$. The Lagrangian can be used to solve for the time development of the generalized coordinates and the corresponding conjugate momenta, $\{p \equiv (\partial \mathcal{L}/\partial \dot{q})\}$:

$$\{p = +(\partial \mathcal{L}/\partial \dot{q}) \; ; \; \dot{p} = (\partial \mathcal{L}/\partial q)\}.$$

These same Lagrangian equations of motion can equally well be expressed in terms of the equivalent Hamiltonian $\mathcal{H}$,

$$\mathcal{L}(q, \dot{q}) \longrightarrow \mathcal{H}(q, p) = K + \Phi.$$

As indicated here, the Hamiltonian usually corresponds to the *total* system energy, kinetic plus potential.

Flows generated by Hamiltonian dynamics are naturally described in $\{q, p\}$ "phase space", rather than coordinate space. Hamilton's equations of motion give time derivatives for both the coordinates and the momenta. The "flow" $\{\dot{q}, \dot{p}\}$ of the variables $\{q, p\}$, is then given in terms of explicit derivatives of the Hamiltonian $\mathcal{H}(q, p)$:

$$\{\dot{q} = +(\partial \mathcal{H}/\partial p) \; ; \; \dot{p} = -(\partial \mathcal{H}/\partial q)\}.$$

The simplest system for which the advantages of generalized coordinates are clear is a particle constrained to circle the origin at a radius of unity.

The motion can be so constrained by using a Lagrangian with a "Lagrange Multiplier" $\Lambda(x, y, \dot{x}, \dot{y})$:

$$\mathcal{L}(x, y, \dot{x}, \dot{y}, \Lambda) = (\dot{x}^2 + \dot{y}^2)/2 + \Lambda(x^2 + y^2 - 1)/2.$$

The resulting equations of motion are $\{\ddot{x} = \Lambda x \; ; \; \ddot{y} = \Lambda y\}$. There are no contributions involving derivatives of Λ because all such contributions are multiplied by the *vanishing* constraint: $(x^2 + y^2 - 1)/2 = 0$. The second time derivative of the constraint,

$$x\ddot{x} + \dot{x}^2 + y\ddot{y} + \dot{y}^2 = 0 = \Lambda(x^2 + y^2) + v^2,$$

can then be solved for the Lagrange multiplier. Λ then takes on the value, $-v^2/r^2$, necessary to maintain the constraint. Exactly the same motion would result under the influence of a proper attractive central force with the magnitude $-v^2/r$. If we choose harmonic forces, with a quadratic potential $r^2/2$, and unit mass, then the corresponding Lagrangian can be written in either Cartesian or polar coordinates:

$$\mathcal{L}(x, y, \dot{x}, \dot{y}) = (\dot{x}^2 + \dot{y}^2 - x^2 - y^2)/2 \longrightarrow$$

$$\{\ddot{x} = \dot{p}_x = -x \; ; \; \ddot{y} = \dot{p}_y = -y\}.$$

$$\mathcal{L}(r, \theta, \dot{r}, \dot{\theta}) = (\dot{r}^2 + r^2\dot{\theta}^2 - r^2)/2 \longrightarrow$$

$$\{\ddot{r} = \dot{p}_r = r\dot{\theta}^2 - r \; ; \; r^2\ddot{\theta} + 2r\dot{r}\dot{\theta} = d(r^2\dot{\theta})/dt = \dot{p}_\theta = 0\},$$

where $p_r \equiv \partial\mathcal{L}/\partial\dot{r} = \dot{r}$ and $p_\theta \equiv \partial\mathcal{L}/\partial\dot{\theta} = r^2\dot{\theta}$. In the polar-coordinate form the conserved nature of the angular momentum, $r^2\dot{\theta}$ is apparent. The two corresponding forms—Cartesian and polar—of the Hamiltonian for a two-dimensional oscillator, follow from the definitions $\{p \equiv \partial L/\partial\dot{q}\}$:

$$\mathcal{H}(x, y, p_x, p_y) = (p_x^2 + p_y^2 + x^2 + y^2)/2 \longrightarrow$$

$$\{\dot{x} = p_x \; ; \; \dot{y} = p_y \; ; \; \dot{p}_x = -x \; ; \; \dot{p}_y = -y\}.$$

$$\mathcal{H}(r, \theta, p_r, p_\theta) = [p_r^2 + (p_\theta/r)^2 + r^2]/2 \longrightarrow$$

$$\{\dot{r} = p_r \; ; \; \dot{\theta} = (p_\theta/r^2) \; ; \; \dot{p}_r = (p_\theta^2/r^3) - r \; ; \; \dot{p}_\theta = 0\}.$$

Evidently the mechanics of a Lagrangian or Hamiltonian system can be described either as a coordinate-space trajectory $\{q(t)\}$ or as a phase-space trajectory $\{q(t), p(t)\}$. The phase-space trajectory view can be generalized to follow the flow of a probability density $f(\{q, p,\}, t)$ through the space. This latter point of view is particularly useful in developing Gibbs' statistical mechanics from Hamiltonian mechanics. As I show in the following Section, Hamilton's flow equations *do* have an important consequence, Liouville's Incompressible Theorem, the best basis for Gibbs' statistical mechanics.

2.5   Liouville's Incompressible Theorem

If we imagine a smoothly-varying probability density $f(q, p, t)$ in the $\{q, p\}$ phase space, with its motion representing collectively all the motions of members of an "ensemble" of systems, all with the same Hamiltonian, but with different initial conditions, it is evident that this phase-space density obeys the "continuity equation",

$$\partial f / \partial t = -\nabla \cdot (fv),$$

where ∇ is a generalized gradient $\sum[(\partial/\partial q) + (\partial/\partial p)]$ and v is the corresponding generalized velocity, with components $\{\dot{q}, \dot{p}\}$. Then, from this "Eulerian" fixed-frame form of Liouville's Theorems, the "comoving", or "Lagrangian", time derivative, following the motion, becomes

$$\dot{f}(q, p, t) = (\partial f / \partial t) + v \cdot \nabla f = -f \nabla \cdot v.$$

This more general "compressible" form of Liouville's Theorem[†] is particularly useful *away* from equilibrium, where the Nosé-Hoover thermostat forces $\{-\zeta p\}$, which we will discuss in Section 2.8, can give rise to a nonvanishing divergence of the flow velocity, $\nabla \cdot v \equiv -\sum \zeta \neq 0$. With Hamilton's equations of motion the divergence vanishes, so that no compressibility can occur:

$$\nabla \cdot v = \sum[(\partial \dot{q}/\partial q) + (\partial \dot{p}/\partial p)] = \sum[(\partial^2 \mathcal{H}/\partial q \partial p) - (\partial^2 \mathcal{H}/\partial p \partial q)] \equiv 0.$$

In the Hamiltonian case the probability density $f(q, p, t)$ flows through the phase space unchanged, just as does the mass density $\rho(r, t)$ of an incom-

[†]See my article on Liouville's Theorems in The Journal of Chemical Physics **109** (1998).

pressible fluid flowing in ordinary three-dimensional space. The corresponding incompressible Liouville Theorem describes the equilibrium situation.
As a corollary, the comoving differential phase-space volume element—
which I denote as $\otimes$—is conserved by Hamilton's equations of motion.

The approximate explicit finite-difference algorithm:

$$\{q_+ = q_0 + (p/m)_0\Delta t \; ; \; (p/m)_+ = (p/m)_0 + (F/m)_+\Delta t\},$$

produces coordinate sequences *identical* to those from the leapfrog algorithm:

$$\{(q_+ - q_0) - (q_0 - q_-) = [(p/m)_0 - (p/m)_-]\Delta t = (F/m)_0(\Delta t)^2\}.$$

It is striking that this finite-difference algorithm satisfies a finite-difference
form of Liouville's incompressible Theorem *exactly*:

$$\otimes_+ = \otimes_0 = \otimes_-.$$

$\otimes$, the *comoving finite* element of phase-space hypervolume, represents what
Gibbs called "extension in phase [space $\{q, p\}$]". It is unknown if the underlying "symplectic property" of the leapfrog algorithm has any significant
effect on the "quality" or the "utility" of the resulting trajectory.

Phase-space flow incompressibility—conserving the "extension in phase"
$\otimes$—is an important defining characteristic of isoenergetic Hamiltonian systems. The incompressible phase-space flow establishes that the corresponding equilibrium phase-space probability density $f(q, p)$ must certainly be
time-independent. Consequently f must have the same *constant* value along
any phase-space trajectory. If there is sufficient *mixing* to cover the entire
energy surface, then necessarily the probability density is constant along
such a surface. The simplest such constant density f defines Gibbs' "microcanonical distribution", or "microcanonical ensemble", which contains
all phase-space states close to a specified energy E,

$$E - (dE/2) < \mathcal{H}(q, p) < E + (dE/2),$$

all with equal weights.

2.6 What *is* Macroscopic Thermodynamics?

From the detailed standpoint of microscopic mechanics an isolated system
is continually undergoing fluctuations and changing state, as is described

by the equations of motion, or, equivalently, by Liouville's incompressible flow theorem. Individual particles can move about. The potential and kinetic parts of the total energy can vary though their sum remains fixed. Rather than considering or even imagining all of these microscopic details for specific phase-space flows it is often worthwhile to ignore them. One can then adopt an alternative *macroscopic* view, analogous to assessing only "average" behavior and focusing on total energy as a *global* "state variable". Thermodynamics takes this macroscopic point of view.

Macroscopic thermodynamics evolved from efforts to design and understand "heat engines"—machines for converting heat into useful work. As a result, heat, work, and energy are among the primitive concepts of thermodynamics. "Thermal variables" needed to describe temperature and heat transfer are also included. In thermodynamics the *efficiency* of macroscopic heat engines converting heat to work is important. Thermodynamic work and heat are analogs of the microscopic potential and kinetic energies. "Work" denotes an energy change in response to *coordinate* variations. Pushing on a piston or lifting a weight are ways of doing work. "Heat" denotes energy changes taking place in the absence of any related coordinate changes. The fraction of the heat taken in which is converted to work *is* the thermodynamic efficiency of a cyclic process. "Reversible" work and heat are abstract thermodynamic concepts having no direct connection to mechanical *time* reversibility. The thermodynamic reversibility concepts instead apply to energy changes or other state changes taking place *through a sequence of equilibrium states*.

Overall, thermodynamics differs from mechanics in three ways. The description of a system's state is *broadened*, to include thermal variables, but *narrowed* through the neglect of fluctuations in the mechanical and thermal variables. Mechanisms for change are omitted too. Thermodynamics describes which states are possible, but does not predict transformation *rates* or *mechanisms*. The thermodynamic description of processes, though including heat flow, is necessarily *macroscopic* and is therefore intrinsically less detailed than the microscopic many-body picture. Despite the limited nature of the thermodynamic view, that view does provide some extremely useful information. It emphasizes what is *possible*. The laws of thermodynamics detailed in the following section establish that it is illegal to have temperature drop in response to added heat, just as it is illegal to have density decrease in response to increasing pressure.

As is discussed more fully in the following section, lacking any knowledge of detailed mechanisms, the *best* one can do, from the standpoint of increasing the thermodynamic efficiency, is to follow an idealized "reversible process". Because the detailed dynamic mechanisms are omitted, thermodynamics is sometimes termed "thermostatics". It is perfectly possible to compute the rate-dependence of irreversible processes by carrying out more detailed simulations incorporating kinetic constitutive information not present in thermodynamics. Simulation of transport processes in macroscopic continua is the subject of Chapter 6.

The wholly new thermal state variables provided by thermodynamics are temperature and entropy. Temperature differences are the driving force behind heat transfer. The integrated effect of reversible heat transfer is entropy change. Entropy, like energy, is a thermodynamic state function. Boltzmann and Gibbs were able to show that the entropy of all those microscopic states linked by a phase-space flow is simply related to the corresponding phase-space hypervolume described in Section 3.5:

$$S_{\text{Gibbs}} \equiv k \ln \Omega.$$

Energy is a well-defined property for any $\{q, p\}$ phase-space state of a microscopic system. Unlike energy, *entropy* is instead a collective, or "ensemble" property, which can be thought of as a system property only through a time-averaging process. Entropy is somewhat subtle. Its value corresponds to the total number of microscopic states consistent with our knowledge. We discuss the connection of the thermodynamic entropy, from integrated heat transfer, to Gibbs' statistical state-counting entropy in Chapter 3.

2.7 First and Second Laws of Thermodynamics

The First Law of Thermodynamics has both global and local versions. These are often stated "The energy of the universe is fixed" or "The energy is a function of state". A system obeying the latter law suffers no net energy change in any cyclic process: $\oint dE = 0$. Thus the First Law prevents any system's acting as a perpetual energy source. The Second Law of Thermodynamics has been formulated in an even greater variety of equivalent ways. Here are four of them: (i) "Heat cannot flow from a colder body to a hotter one"; (ii) "Entropy must increase"; (iii) "No *cyclic* process can convert heat entirely to work"; (iv) "In any *cyclic* process the heat Q transferred *to* the system *from* its surroundings at the temperature T must

obey an *inequality:* $\oint dQ/T < 0$." These four statements, though equivalent from a macroscopic point of view, are not equally useful on the more
detailed microscopic level. The first of these statements is inconsistent with
the existence of fluctuations—despite the unquestioned fact that heat conductivity is positive, it is always possible, for sufficiently small temperature
gradients in sufficiently small volumes, to find heat traveling in the *wrong*
direction, for a while. The second of the Second Law statements is likewise
inconsistent with a time-reversible dynamics. Leaving aside the problem
of *defining* nonequilibrium entropies, any situation in which the entropy
"correctly" increases could be made "incorrect" by reversing the velocities.
The last two statements, dealing with *cyclic* processes, are more promising.
Long time averages, over *many* cycles, *can* be thermodynamically-correct,
both microscopically and macroscopically.

The Second Law of Thermodynamics, stated as a long-time average, either over a large number of cycles or for a stationary state, need not conflict
with microscopic mechanics, once that mechanics is generalized to include
the concepts of heat and temperature. The cyclic-process form of the Second Law does *not* apply to isolated systems, which are both artificial and
uninteresting. It is evident that, with rather mild assumptions (that the accessible phase space is bounded), an isolated system will eventually return
arbitrarily closely to its initial state ("Poincaré recurrence"). It is also evident that any evolution thought to correspond to an entropy increase would
necessarily correspond also, when reversed, to a compensating entropy decrease ("Loschmidt's paradox"). Thus it is unreasonable, on both counts,
to expect an isolated system to show a systematic entropy increase. But
the time-averaged version of the Second Law *is* true, $\langle \oint dQ/T \rangle < 0$. This
inequality states that cyclic processes generate entropy, *provided* that an
average over many repetitions of the cycle is implied. In this time-averaged
form the Second Law of Thermodynamics can be *proven*[‡] as a theorem in
thermomechanics. It applies to simple prototypical stationary processes,
like shear flow and heat flow, as well as to more complex flows.

The most interesting, puzzling, and enduring aspect of time reversibility
is the connection between atomistic microscopic dynamics and the macroscopic Second Law of Thermodynamics. Most explanations for this connection, though without doubt correct, are somewhat limited in scope.
Boltzmann's dilute-gas approach, together with Loschmidt's and Zermélo's

---

[‡]See page 269 in my *Computational Statistical Mechanics*, as well as Section 7.8 here.

objections to it, mentioned in Chapter 1, is the most familiar. Green and Kubo's much more recent linear-response theory of transport is a first-order perturbation theory. It applies to liquids and solids as well as to gases. Objections to Green-Kubo theory are less convincing and not so well organized as those Boltzmann faced. For more mathematical discussions, see the reviews of Zwanzig and Ichiyanagi. Perhaps these objections are also less important, because the Green-Kubo-Onsager approach is limited to states which are close to equilibrium and because the results of the theory can be obtained in *many* alternative ways.

It might well be thought that time-reversible equations of motion are inconsistent with the symmetry-breaking inherent in the dissipative shrinking flows called "strange attractors" that characterize nonequilibrium problems[§]. The ways in which time-reversible dissipation can, and does, occur have become clear with the development and analysis of computer algorithms, particularly during the past fifteen years. These algorithms, and related simulations, represent and elucidate new paths to understanding the reversibility paradox faced earlier by Boltzmann, Green, and Kubo. In order to describe the new paths we first must introduce temperature, and thermostats, into the microscopic equations of motion. The distribution functions which then result are *fractal* distributions, which lead directly to an understanding of nonequilibrium steady states and the time-averaged micromechanical version of the Second Law.

2.8   Temperature, Zeroth Law, Reservoirs, and Thermostats

In order to simulate processes involving thermodynamic work and heat transfer it is useful to adopt a microscopic *definition* of temperature, corresponding to the temperature measured by an ideal-gas thermometer. The usefulness of temperature as a state variable can be stated as the Zeroth Law of Thermodynamics, "Two bodies in thermal equilibrium with a third are also in thermal equilibrium with each other", meaning that there is no net transfer of heat between any two bodies with the same temperature T. This macroscopic thermodynamic concept ignores fluctuations. It is certainly false, on a local and instantaneous level. But an equivalent microscopic concept, for stationary boundary conditions, *can* be expressed as

---

[§]It *is* "strange" that an exponentially unstable flow *shrinks*, on average: $\langle d \ln \otimes / dt \rangle < 0$.

a time average: "Two bodies with no long-time tendency to transfer heat with a third have the same time-averaged temperature, $\langle T \rangle$".

Such a microscopic zeroth law provides no operational definition for the equilibrium temperature common to the three bodies. Among many possibilities, by far the most natural, and most convenient microscopic definition of T, or $\langle T \rangle$, is the classical ideal-gas temperature, given in terms of the (time-averaged) mean-squared velocity: $kT_{xx} \equiv \langle mv_x^2 \rangle$. In a stationary equilibrium situation it is not necessary to distinguish between T and $\langle T \rangle$. Likewise, at equilibrium T_{xx} and T_{yy} are equal. This temperature definition follows automatically if one considers the "third body" of the zeroth law to *be* an ideal gas thermometer, with enough degrees of freedom to make the fluctuating difference, $T - \langle T \rangle$, negligible[¶].

Such a thermometer is most simply conceived of as containing many infinitesimal particles. Their interaction is very weak, $\Phi \simeq 0$, but strong enough to provide the equilibrium Maxwell-Boltzmann distribution proportional to $e^{-mv^2/2kT}$ and to establish the isotropic ideal gas law within the thermometer:

$$P_{xx}V = P_{yy}V = P_{zz}V = NkT.$$

With the ideal-gas temperature scale defined, it is possible to model both equilibrium and nonequilibrium thermostats in either of two ways: by constraining the instantaneous kinetic energy or by constraining its time average. The *first* choice gives an "isokinetic" thermostat. The thermostat has to be implemented by imposing an additional thermostat force, which constrains the kinetic energy to a fixed value. The potential energy is left free to fluctuate. For a sufficiently mixing system this approach leads to Gibbs' canonical distribution for the potential energy, $\propto e^{-\Phi/kT}$. Finding the required constraint force is a straightforward application of Gauss' "Principle of Least Constraint" or the closely-related variational principle considered by Gibbs[‖].

Gauss' Principle states that any required constraint forces $\{F_c\}$ are to be chosen so as to *minimize* the corresponding sums, over the constrained degrees of freedom: $\{\sum F_c^2/2m\}$. By applying Gauss' Principle to the constraint of fixed kinetic energy, $\dot{K} = 0$, an additional "frictional force" results, $-\zeta p$ for each thermostated degree of freedom. The friction coeffi-

---

[¶] A detailed kinetic treatment of the ideal-gas thermometer appears in the next Section.

[‖] See Gibbs' 1879 paper, reproduced in full as Appendix V of L. P. Wheeler's biography.

cient ζ can be chosen to keep the kinetic energy constant. For an otherwise isolated system, with equations of motion $\{\dot{p} = F - \zeta p\}$, it is evident that the choice $\zeta \equiv \sum F \cdot (p/m)/2K$ implies a fixed kinetic energy, $K = K_0$.

The resulting isokinetic dynamical system is not among the usual types represented by Gibbs' ensembles. More usual is the canonical case corresponding to thermal equilibrium at a temperature T, with the phase-space density $f(q, p) \propto e^{(A-\mathcal{H})/kT}$, where Helmholtz' free energy $A \equiv E - TS$ provides a normalization for the "Boltzmann factor" $e^{-\mathcal{H}/kT}$ in Gibbs' formulation of the canonical ensemble and its "partition function" Z:

$$(1/N!)\prod[\int \int dq dp/h]e^{-\mathcal{H}/kT} \equiv e^{-A/kT} = e^{+S/k}e^{-E/kT} = Z(N, V, T).$$

In 1984 Shuichi Nosé developed a more general approach to thermostated mechanics. Fluctuations were included, corresponding to the *second* choice of thermostat, one in which the *time average* of the ideal-gas thermometer temperature, $\langle T \rangle = \langle mv_x^2/k \rangle$, is constrained. In his mechanics the instantaneous temperature is allowed to fluctuate, in such a way as to fill out the entire canonical distribution *if* the dynamics is sufficiently mixing. Nosé's original formulation included an extraneous "time-scaling" variable s.[**] After meeting with Nosé and studying his work, I suggested that s is best omitted. The resulting simpler formulation, free of time scaling or effective masses, is usually called "Nosé-Hoover mechanics". Nosé-Hoover mechanics includes additional constraint forces, $\{-\zeta p\}$ ensuring that the long-time-averaged kinetic energy $\langle K \rangle$ approaches the canonical equipartition value from Gibbs' statistical mechanics:

$$\langle K \rangle_{t \to \infty} \equiv K_0 = DNkT/2,$$

where T is the temperature and D is the dimensionality of the system. The Nosé-Hoover feedback forces $\{-\zeta p\}$ include a characteristic thermostat relaxation time τ. For an otherwise isolated system, the equations of motion take the form of "integral control" equations, with the friction coefficient proportional to the time-integrated deviation of K from K_0:

$$\{\dot{q} = (p/m) \; ; \; \dot{p} = F - \zeta p \; ; \; \dot{\zeta} = [(K/K_0) - 1]/\tau^2.$$

Taking the long-time average of the last equation, with the assumption that the mean value of the friction coefficient $\langle \zeta \rangle$ is bounded, establishes that

[**]This extra variable could equally well be thought of as representing an effective mass.

the mean value of the kinetic energy must approach the constant K_0:

$$\langle \zeta \rangle \text{ constant} \longrightarrow \langle \dot{\zeta} \rangle = 0 \propto \langle [(K/K_0) - 1] \rangle \longrightarrow \langle K \rangle = K_0.$$

Thus the Nosé-Hoover approach (i) fixes the *time-averaged* kinetic energy, rather than the instantaneous value and (ii) introduces a useful phenomenological relaxation time τ with which to simulate *rates* of heat transfer.

At equilibrium these motion equations are exactly consistent with Gibbs' canonical distribution. That is, with the *assumption* that f has the canonical form, the Nosé-Hoover motion equations imply that it is stationary:

$$f(q, p, t) \propto e^{-\mathcal{H}/kT} \longrightarrow (\partial f/\partial t)_{\text{Nosé-Hoover}} \equiv 0 \longrightarrow f = f(q, p).$$

Nosé-Hoover mechanics is the simplest *dynamical* analog of Gibbs' statistical mechanics. It is an example of "thermomechanics", an augmentation of Newtonian mechanics which includes temperature control for selected degrees of freedom. Though his original derivation was unnecessarily complicated, with "time scaling" included, Nosé's idea led to something new and useful—a time-reversible dynamics in which the phase-space probability density f changes in response to heat transfer, $d \ln f/dt \equiv -\nabla \cdot v = \sum \zeta$, as was discussed in Section 2.5.

There is another elegant way to avoid Nosé's "time scaling", while retaining the flexibility of the relaxation time τ, the physical interpretation of phase-space compressibility, and an exact link to Hamiltonian mechanics. This approach was discovered by Carl Dettmann, in 1996. For a system with # thermostated degrees of freedom, Dettmann showed that the Nosé-Hoover equations of motion also follow naturally when a special Hamiltonian, resembling that which Nosé had used, is set equal to zero:

$$\mathcal{H}_{\text{Dettmann}} = s\mathcal{H}_{\text{Nosé}} = \sum (p^2/2ms) + s[\Phi + \#kT \ln s + (p_s^2/2M)] \equiv 0.$$

In the absence of the special choice $\mathcal{H} \equiv 0$ Dettmann's Hamiltonian bears a superficial resemblance to Nosé's, in which s is a time-scaling variable. But Dettmann's s is simply a new "thermostat variable" (its logarithm is proportional to the time integral of ζ). Dettmann's approach completely avoids the need for any scaling of time. His motion equations are:

$$\{\dot{q} = (p/ms) ; \dot{p} = sF\},$$

$$\dot{s} = s(p_s/M);$$

$$\dot{p}_s = -\partial \mathcal{H}/\partial s = \sum(p^2/2ms^2) - [\Phi + \#kT\ln s + (p_s^2/2M)] - \#kT.$$

Evidently the essential constraint $\mathcal{H} = 0$ allows the uninteresting variable s to be omitted and permits the (q, p, ζ) phase-volume to change with time, despite the Hamiltonian basis. Further, the arbitrary restriction $\mathcal{H} = 0$ makes it possible to simplify the evolution equation for the friction coefficient: $p_s \propto \zeta$:

$$\dot{p}_s = \sum(p^2/ms^2) - \#kT \ ; \ \zeta = (p_s/M) = \int_0^t [(K/K_0) - 1]dt'/\tau^2.$$

If we abbreviate the combination $(p/ms) \equiv v$, for velocity, the complete scheme reduces exactly to the Nosé-Hoover equations:

$$\{\dot{q} = v \ ; \ \dot{v} = (F/m) - \zeta v\} \ ; \ \dot{\zeta} = \sum[(mv^2/kT) - 1]/\tau^2.$$

Although this approach sometimes fails to generate the entire canonical density distribution, relatively simple modifications, illustrated on page 69, can accomplish this. Frictional forces, of the form $-\zeta p$, are common to *all* computationally-useful deterministic approaches to thermomechanics. Gauss' isokinetic form also corresponds to the simple idea of velocity rescaling used in early computer simulations. Gauss' thermostat is also the instantaneous $\tau \longrightarrow 0$ limit of Nosé-Hoover mechanics.

2.9 Irreversibility from Stochastic Irreversible Equations

A longstanding route to simulating irreversible behavior is the Langevin equation of motion for a "heavy" or "Brownian" particle of mass M,

$$M\ddot{r} = \dot{p} = F - (p/\tau) + F_{\text{stochastic}}.$$

The characteristic relaxation time τ is a constant and the random numbers underlying the "stochastic" force are to be chosen so as to recover the desired temperature. This approach contains two irreversible ingredients: the drag force $(-p/\tau)$, which if reversed would lead to exponentially-rapid divergence of the kinetic energy, and the stochastic forces, which introduce a stream of "information" into the dynamics through the underlying random numbers. Although both sources of irreversibility seem to be far from fundamental physics, it is not difficult to "derive" such an irreversible equation—with a drag force—by carrying out an *average* over collisions. For simplicity, I demonstrate this for a one-dimensional problem.

Consider a large heavy particle with mass M and velocity V undergoing a collision with a small light particle representative of a "heat reservoir": an ideal-gas thermometer made up of particles with mass m. The velocity v is chosen from an equilibrium Maxwell-Boltzmann distribution. *Heat reservoir* particles are *always* at equilibrium. In the center-of-mass frame, the velocity of the heavy particle necessarily changes sign in such a way as to conserve momentum and energy:

$$V - (MV + mv)/(M + m) \longrightarrow (MV + mv)/(M + m) - V.$$

We can analyze collisional effects systematically, as series in the square root of the mass ratio (m/M). It is necessary to keep in mind that the equilibrium ratio of v/V is of order $(M/m)^{1/2}$. Thus the two velocities in the "first term" of an explicit expansion in powers of (m/M),

$$\Delta V = -2(V - v)(m/M),$$

actually differ, on the average and near equilibrium, by a factor $(m/M)^{1/2}$.

To begin, we calculate the average effect of light-particle collisions with the heavy particle. Choose an ideal-gas number density n with a thermal distribution of velocities $\{v\}$ corresponding to the temperature T. Because the heavy-particle collision rate is proportional to the relative speed, the average requires a collision probability proportional to the relative velocity, $V - v$. The mean collision rate $\langle \Gamma \rangle$ follows directly from the Maxwell-Boltzmann distribution. Assuming that the massive particle velocity V is negligibly small, the result is

$$\langle \Gamma \rangle = n(m/2\pi kT)^{1/2} \int_{-\infty}^{+\infty} |v - V| e^{-mv^2/2kT} dv \longrightarrow$$

$$n(m/2\pi kT)^{1/2} \int_{-\infty}^{+\infty} |v| e^{-mv^2/2kT} dv = n(2kT/\pi m)^{1/2},$$

where the integral ranges over all values of v so as to include collisions from both left and right.

The calculation of the *averaged* heavy-particle momentum change,

$$\langle V \rangle \longrightarrow \langle V + 2(v - V)[m/(M + m)] \rangle,$$

proceeds in a similar way. To first order in the mass ratio (m/M) the mean rate at which the velocity changes is

$$\langle dV/dt \rangle \longrightarrow \langle 2\Gamma(v-V)(m/M) \rangle =$$

$$n(m/2\pi kT)^{1/2} \int_{-\infty}^{+\infty} 2(v-V)(m/M)|v-V|e^{-mv^2/2kT}dv.$$

Introducing the relative velocity, $\alpha = v-V$, the integral can be rewritten as an integral over α, and the exponent can be expanded:

$$e^{-mv^2/2kT} = e^{-m\alpha^2/2kT}[1 - (m\alpha V/kT) + \dots].$$

The first term in the square brackets does not contribute to the integral, which is odd in α; the second term gives the final result:

$$\langle \dot{V} \rangle \to n(m/2\pi kT)^{1/2} \int_0^\infty \frac{-4\alpha^3 m^2 V}{MkT} e^{-m\alpha^2/2kT}d\alpha = -4\langle\Gamma\rangle(m/M)V.$$

The averaged effect is a friction coefficient $4\langle\Gamma\rangle(m/M)$. These straightforward calculations can be carried out in two or three dimensions with similar results. They also show that collisions with the bath particles provide a frictional damping force, proportional to the heavy particle velocity V and the mean collision rate $\langle\Gamma\rangle$.

The mean *energy* change for the heavy particle, due to the bath collisions, can be calculated in a similar way. Expressing the *energy* change in terms of the relative velocity, $\alpha = v - V$, gives terms both linear and quadratic in α:

$$\langle \dot{E} \rangle = (M/2)\langle[4\alpha Vm/(M+m)] + [4\alpha^2 m^2/(M+m)^2]\rangle,$$

where the average on the righthand side is over collisions. Using again an expansion of $e^{-mv^2/2kT}$, and keeping only terms of order (m/M) the mean energy change rate becomes

$$\langle \dot{E} \rangle = 4\langle\Gamma\rangle(m/M)[kT - MV^2].$$

This extremely interesting result shows that the kinetic energy of the massive particle relaxes smoothly toward $kT/2$. This relaxation is quite general. It is easy to show, for a two-dimensional hard disk or a three-dimensional hard sphere, interacting with a two- or three-dimensional ideal-gas thermometer, that the kinetic energy of the hard particle, $MV^2/2$, relaxes toward kT or $3kT/2$.

The argument just given closely resembles the statistical arguments used by Boltzmann in deriving the Boltzmann equation. In both cases irreversible behavior is the result of statistical averaging over the completed collisions in an underlying time-reversible mechanics. This same irreversible relaxation can also be represented by a much less cumbersome model, the phenomenological Langevin equation. Here, the viscous drag force, plus a "stochastic" noise force sufficient to maintain the kinetic energy are added to the single-particle equation of motion:

$$\dot{p} = F - (p/\tau) + F_{\text{stochastic}}.$$

The stochastic noise force is usually chosen from an appropriate Gaussian distribution. Klages, Rateitschak, and Nicolis have recently considered the use of deterministic time-reversible maps as thermostat forces. More realistic forms could be chosen provided they incorporate many oscillations during the decay time of the drag force, $\tau = 1/\zeta$. The resulting heavy-particle "temperature" will then reproduce that of the bath. In some cases it is desirable to consider a very low-temperature bath, giving the simpler *noise-free* motion equation $\dot{p} = F - \zeta p$. Provided that the force F represents a *driven* system such a low-temperature bath leads to a simple nonequilibrium steady state with $T_{\text{system}} >> T_{\text{bath}}$. It has been shown that this low-temperature limit of the Langevin equation leads to the fractal attractors usually associated with dissipative irreversible flows. See my 1999 paper on "Steady States" in Physics Letters A.

2.10 Irreversibility from Time-Reversible Equations?

In the remainder of this book, I discuss links between the microscopic time-reversibility of atomistic dynamics and the macroscopic thermodynamic irreversibility described by the Second Law of Thermodynamics. From the computational standpoint, the simplest way to forge such a link makes use of the mechanical ideal-gas temperature concept, with heat reservoirs based on Gauss' or Nosé-Hoover mechanics. This approach leads to irreversible dissipative solutions from time-reversible equations of motion.

Temperature can easily be introduced into computer simulations by using special forces, of the kind developed by Nosé. In principle, the special forces could have been avoided, at least temporarily, by instead studying "open systems" attached to large reservoirs of mass, momentum, and energy. However, the sources and sinks which drive open systems must ultimately be given operational significance too, and their own thermostats. It is plainly better to honor Occam, taking reservoirs with a few degrees of freedom rather than many. For *linear* transport processes it can be confidently expected that the values for closed systems and open systems will agree[††]. This is a requirement for any useful computer algorithm.

We will see that solutions following the *time-reversible* friction-coefficient approach share many features with solutions of *irreversible* dissipative equations of motion. Both approaches produce nonequilibrium multifractal attractor states with a family resemblance to the Lorenz attractor of Section 4.9.1 and to the Sinai-Ruelle-Bowen states mentioned in Section 8.10. Some of the formal difficulties encountered in describing nonequilibrium flows with multifractal distributions are fundamental. It is hard to imagine a useful mathematical description of a fractal flow through phase space, particularly in the case that the initial distribution is fractal too. Formally, this apparent difficulty can only be avoided by considering coarse-grained distributions, reflecting the limited resolution, or information content, of any conceivable measurement or simulation.

2.11 Example Problems

The first example considered here is the two-dimensional *dissipative* Baker Map. This is an instructive caricature of flows which incorporate a changing phase volume, $\langle \dot{\otimes} \rangle \neq 0$. Such flows obey the local *compressible* form of Liouville's Theorem for the time-development of f and an *infinitesimal* extension in phase, the comoving hypervolume $\otimes$:

$$\dot{f}/f \equiv -\dot{\otimes}/\otimes = -\nabla \cdot v.$$

Compare the singular structure which the dissipative map generates to the smooth and continuous structure from the equilibrium *incompressible* Baker Map considered in Chapter 1. The second example is a continuous flow,

[††]See Evans and Morriss' *Statistical Mechanics of Nonequilibrium Liquids*.

in two space dimensions, to facilitate visualization, but complex enough to reproduce Gibbs' canonical distribution. It is the motion of a mass-point particle through a periodic two-dimensional array of soft scatterers.

2.11.1   *Time-Reversible Dissipative Map*

In Chapter 1 we considered a caricature of equilibrium systems, with incompressible phase-space flows, $\dot{\otimes} = 0$. This was an area-conserving Baker Map B with a stationary ergodic solution $f = 1/4$ within a 2×2 square. Here we consider a "dissipative" modification D of that map. The dissipative map, $D_{xy}(x,y) = (x', y')$ or the rotated version $D_{qp}(q,p) = (q', p')$ is *compressible*, with $\langle \dot{\otimes} \rangle \neq 0$. The upper leftmost *third* of the same 2×2 square is mapped into the upper rightmost *two thirds* of that square:

$$+1/3 < y < +1 \longrightarrow \{x,y\}' = \{(2x+1)/3, (3y-2)\};$$

$$-1 < y < +1/3 \longrightarrow \{x,y\}' = \{(x-2)/3, (3y+1)/2\}.$$

Written in terms of a (horizontal) "coordinate" $q \equiv (x-y)/\sqrt{2}$ and a (vertical) "momentum" $p \equiv (x+y)/\sqrt{2}$, within the square $|\pm q \pm p| < \sqrt{2}$, the rotated map D_{qp} becomes:

$$q < (p - \sqrt{2/9}) \longrightarrow q' = (11q/6) - (7p/6) + \sqrt{49/18},$$

$$p' = (11p/6) - (7q/6) - \sqrt{25/18};$$

$$(p - \sqrt{2/9}) < q \longrightarrow q' = (11q/12) - (7p/12) - \sqrt{49/72},$$

$$p' = (11p/12) - (7q/12) - \sqrt{1/72}.$$

This time-reversible but dissipative map produces ergodic steady-state solutions which are distinctly different to those of the Equilibrium Baker Map of Chapter 1. The *dissipative* map includes area *changes*, leading to a *fractal* distribution. The distribution is actually "multifractal", meaning that the singular power-law dependence of density on distance varies with position. More details of such chaotic systems are given in Chapter 7. This problem is a useful introduction to dissipative chaos. The map is termed "dissipative" meaning that information is *lost* in the contraction of the mapping in the stable x direction. The simultaneous spreading, in the *unstable y* direction, smoothes and destroys any irregularities in that direction, and *creates* information, with formerly insignificant differences governing present and future behavior.

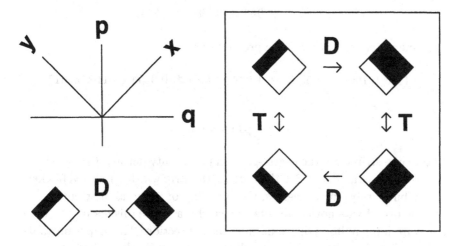

Figure 2.3: The Dissipative Baker Map, in (x, y) coordinates, is shown at the left. The rotated map, in (q, p) coordinates, is shown at the right. The mapping operation is denoted D; the *time-reversal* operation T corresponds to changing the sign of the "momentum" p with the coordinate q fixed.

This dissipative map includes periodic cycles analogous to those found in solving the equilibrium Baker Map. The two-iteration cycle links the points $(x, y) = (-\frac{5}{7}, +\frac{5}{7})$ and $(x, y) = (-\frac{1}{7}, +\frac{1}{7})$. These correspond to the *rotated* cycle linking the points $(q, p) = (-\frac{5}{7}\sqrt{2}, 0)$ and $(q, p) = (-\frac{1}{7}\sqrt{2}, 0)$.

The inhomogeneous probability density can be characterized through a "coarse-grained" evaluation of an "entropy", $-k\langle \ln p \rangle$. The probability p is the product of the box area and the probability density f. This coarse-grained entropy diverges as the box width is decreased. Because the stretching in the y direction is Lyapunov unstable, eventually leading to a smooth, constant probability density in that direction, it is evident that the two strips $\{-1 < x < -1/3, -1/3 < x < +1\}$ are eventually occupied with probabilities of 2/3 and 1/3. A numerical calculation, with fifty million points, gave probabilities for occupying three strips of width 2/3 which were consistent with this expectation:

$$\{p\} = \{0.66664, 0.29512, 0.03817\} \longrightarrow \sum - p\ln p = 0.6874;$$

$$\{f\} = \{0.99996, 0.44268, 0.05725\}.$$

For nine strips of width 2/9 the probabilities were:

$$\{0.4443, 0.1968, 0.0255, 0.1968, 0.0255, 0.0729, 0.0254, 0.0110, 0.0017\}$$

$$\longrightarrow \sum - p\ln p = 0.6973.$$

In the nine-strip case, the coarse-grained probability density f ranges from 2, in the first strip, to 0.008 in the last. As the strip width is reduced further, slowly but surely $\langle \ln p \rangle$ diverges, the signature of a fractal distribution.

The two "Lyapunov exponents" which characterize this dissipative map follow from the uniform distribution in the y direction. The map is unstable in that direction, expanding a small perturbation Δy by a factor 3 with probability 1/3, and by a factor 3/2 with probability 2/3. Thus the larger exponent is

$$\lambda_1 = (1/3)\ln(3/1) + (2/3)\ln(3/2) = +0.63651.$$

Similarly, in the stable x direction Δx shrinks by a factor 3 with probability 2/3 and by a factor 3/2 with probability 1/3, leading to

$$\lambda_2 = (2/3)\ln(1/3) + (1/3)\ln(2/3) = -0.86756.$$

Unlike the equilibrium Baker Map the stationary solution of this dissipative (but time-reversible) Baker Map is necessarily *singular everywhere*. This is because repeated iteration of the map continually magnifies the separation of nearby points in the y direction, eventually leading to a bifurcation between y coordinates which are greater than 1/3 and those which are less. The result is the fractal distribution shown in Figure 2.4. This distribution, despite its ergodic behavior, has an "information dimension"—described below—of only 1.734, significantly less than that of the space in which the distribution is embedded.

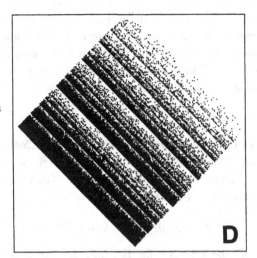

Figure 2.4: 100,000 points from the time-reversible dissipative Baker Map shown in Figure 2.3. Evaluation of the "information dimension" for this multifractal distribution gives $D_I = 1.734$.

The entire 2×2 square constitutes an *attractor* with two thirds of the square (y values less than $1/3$) contracting and the remaining third expanding. Any sequence of points $\{(\pm q, +p)\}$ produced by the mapping can *formally* be converted to another satisfactory sequence $\{(\pm q, -p)\}$ by reversing the *order* of the points as well as the sign of p. Reversing the sequence of the points changes the signs of both Lyapunov exponents (by replacing the expanding and contracting processes with their inverses) so that the sum,

$$(\lambda_1 + \lambda_2)_{\text{reversed}} = 0.86756 - 0.63651 = 0.23105,$$

is positive, corresponding to an unstable—and therefore unobservable—repellor. We will see that this same structure, attractor-repellor pairs, occurs in time-reversible many-body systems, away from equilibrium. The *inevitable* contracting attractor states obey the Second Law of Thermodynamics. The *unobservable* repellor states would violate that Law.

The dimensionality of fractal distributions can be characterized in a variety of ways. The "box-counting" dimension and the "information dimension" are the most useful. The *box-counting* dimension gives the limiting power-law dependence of the number of occupied boxes # on the (sufficiently small) box size ϵ :

$$D_{\text{BC}} \equiv \ln \# / \ln(1/\epsilon).$$

This time-reversible dissipative Baker Map is *ergodic* and, with enough data, eventually occupies *every* square box, so that the box-counting dimension of the fractal is exactly 2. The *information* dimension describes the limiting power-law dependence of the box probabilities, $\langle -\ln p \rangle$, on the box size. Using 3^n square boxes of size $\{\epsilon = 2(1/3)^n\}$ provides a series of estimates for the information dimension, as n increases and ϵ decreases:

$$n = 1 \longrightarrow 1.6874, n = 2 \longrightarrow 1.6973, \ldots, n = \infty \longrightarrow 1.734.$$

The Gibbs' entropy, $S_G/k \equiv -\langle \ln f \rangle \simeq 0.266 \ln \epsilon$, associated with the Baker Map *diverges* as the box size is reduced. See Figure 2.5.

The power-law divergence of the Gibbs entropy is closely related to the information dimension of the fractal:

$$D_I = \langle \ln p \rangle / \ln \epsilon = \langle (\ln f \epsilon^2) \rangle / \ln \epsilon = 1.734 - 0.06(1/n)^{0.20}.$$

Thus the estimate for the information dimension of the two-dimensional dissipative Baker Map is 1.734. This corresponds to the small-ϵ divergence of the probability density as follows:

$$f_{\epsilon \to 0} \propto \epsilon^{1.734 - 2.000}.$$

This divergence of f indicates that "almost all" of the probability is found in a "negligibly small" fraction of the boxes, the attractor "core". For many interesting illustrations, and a clear overview of the geometrical significance of attractor dimensions, see the 1983 paper by Farmer, Ott, and Yorke. Schröder's is the most helpful book I have found in this area.

Figure 2.5: Coarse-grained Gibbs entropy, relative to that of the equilibrium Baker Map. The data are 40,000,000 points, which show divergence of $\langle \ln f \rangle$, varying as $0.266 \ln(1/\epsilon)$ for $3^{-13} < \epsilon < 3^{-1}$.

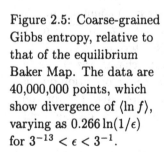
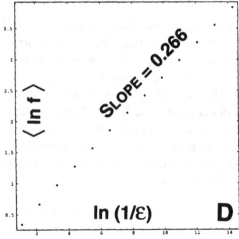

Kaplan and Yorke pointed out that the information dimension should correspond to that linearly-interpolated dimension at which the partial Lyapunov-exponent sum *changes sign*. With only two Lyapunov exponents the one-term "sum" is $\lambda_1 > 0$ while the two-term sum is $\lambda_1 + \lambda_2 < 0$. The "Kaplan-Yorke dimension" is $1 + |\lambda_1|/|\lambda_2|$. For the Baker Map the information and Kaplan-Yorke dimensions are *identical*:

$$D_{\mathrm{I}} = D_{\mathrm{KY}} = 1 - (\lambda_1/\lambda_2) = 1 - (+0.63651/-0.86756) = 1.73368,$$

as is known to hold exactly in this case[‡‡].

Similar singular fractal characteristics are shared by the more complicated nonequilibrium many-body phase-space distributions. The probability density typically diverges, indicating the concentration of the probability onto a fractal set, but with a box-counting dimension characteristic of equilibrium. The simple Baker Map example is a useful caricature of real flows. The analysis of many-body flows relies on the Kaplan-Yorke conjecture because binning and box-counting operations are impractical in spaces of more than a few dimensions.

2.11.2 *Time-Reversible Smooth Galton Board*

To simplify the required computer programming, let us consider a dynamic chaotic problem with smoothly varying, rather than singular, forces. Let a single unit-mass point particle move in the periodic potential generated by a two-dimensional lattice of *soft* scatterers. By choosing a very smooth force law for the scattering, derived from a short-ranged pair potential with three vanishing derivatives at the cutoff distance $r = 1$:

$$\phi(r < 1) = 3(1 - r)^4,$$

it is relatively easy to generate trajectories covering *millions* of scattering collisions. To ensure the time-reversibility of the trajectories, the Størmer-Verlet "leapfrog" algorithm is the simplest choice:

$$\{r_+ - 2r_0 + r_- = \Delta t^2 (F/m)_0\}.$$

For energies of order unity, an error analysis of the algorithm suggests, as simulations bear out, that timesteps in the range $0.005 \leq \Delta t \leq 0.05$ give reasonably accurate accelerations over run lengths on the order of a million

[‡‡]See Dorfman's book for more details and discussion, with a reference to the proof.

collisions. (The "local" single timestep coordinate error is of order Δt^4 while the global error, at a fixed time, is *formally* of order $t\Delta t^2$, but *actually* of order $e^{\lambda t}$, due to Lyapunov instability. See page 11 of *Computational Statistical Mechanics*.)

As Levesque and Verlet pointed out, the integer version of this algorithm provides strict time reversibility. Such a problem has, in principle, a periodic solution (provided that the algorithm does not, eventually, diverge) but presentday computers are still not fast enough for a straightforward investigation of the periodicity. Numerical work shows that at least eight-digit accuracy is required, for which the determination of Poincaré recurrence periods of length $\simeq 10^{16}\Delta t$ is an unappealing prospect. Probability densities, and the comparison of dynamical averages with Gibbs' ensemble theory, suggests that this problem is *ergodic* (covering all the available energy states) over a wide range of energies.

Using a periodic 2×2 square unit cell, and a moving particle of unit mass, total energies less than 3 appear to provide an ergodic coverage of the available phase space. See the discussion of numerical ergodicity in the next Chapter. The values of the time averages $\langle \Phi = \sum \phi \rangle$ and $\langle \Phi^2 \rangle$ support the apparent ergodicity, agreeing well with the predictions of Gibbs' isoenergetic microcanonical-ensemble theory for this problem.

2.12   Summary

The exact time-reversibility found in the classical equations of motion for isolated systems can be retained in thermomechanical computer simulations of systems which interact with their environments. Thermal environments can be characterized by time-reversible "thermostats" constraining the kinetic energy of selected degrees of freedom. A variety of time-reversible approaches to thermostats all lead to forces linear in the momenta, $\{-\zeta p\}$, as does also the low-temperature limit of the *irreversible* stochastic approach summarized by the Langevin equation. In all of the time-reversible approaches to thermomechanics, the friction coefficient ζ changes sign along any time-reversed trajectory.

Computer simulation of chaotic trajectories necessarily requires a discrete truncated caricature of hypothetical continuous trajectories. In principle, the discrete and finite nature of the numerical state space would seem to make the dynamics vulnerable to the objections raised by Loschmidt and Zermélo. Although these objections are valid for any isolated system, adding an interaction with the surroundings provides a simple geometric understanding of a (time-averaged) Second Law of Thermodynamics.

The presence of chaos, or "sensitivity to initial conditions" does not itself affect trajectory reversibility, but, with the added influence of a system's surroundings, provides a new mechanism for irreversible behavior. This mechanism is the formation of fractal attractors with an information dimension which decreases as the departure from equilibrium increases. The combination of dissipation with time-reversibility yields strange attractors with a family resemblance to the fractal distribution generated by the dissipative Baker Map. The association of dissipation with fractals will be described in greater detail in Chapters 7 and 8.

Chapter 3

Gibbs' Statistical Mechanics

Our life is frittered away by detail ...
Simplify, simplify.

Henry David Thoreau

3.1 Introduction

Energy is *the* basic state function in thermodynamics. For a given amount of material in a particular state—(V, T) or (P, T) or (P, V), for instance—the equilibrium energy is always the same. *Changes* of energy in mechanics involve work, the product of force and displacement. In thermodynamics the work done by a fluid expanding reversibly against a pressure P is the integral $\Delta W = \int P dV$. Thermodynamics describes not only such energy changes due to mechanical work, but also those due to heat transfer. A thermodynamic description of heat transfer requires the *two* additional state functions discussed in Chapter 2, temperature—defined in terms of the ideal-gas thermometer—and entropy S. The *reversible* transfer of heat *to* a fluid *from* a heat reservoir at temperature T is the integral $\Delta Q \equiv \int T dS$. Thus thermodynamic entropy is defined by an integration linking an initial standard state to any other current state of interest. The integration must follow a thermodynamically *reversible* process:

$$\Delta S(N, E, V) \equiv \int_{\text{rev}} (1/T) dQ.$$

Though this thermodynamic approach to mechanical and thermal energy changes gives the appearance of generality, sufficiently complex systems frustrate the need for an underlying reversible process. Thermodynamically reversible processes must take place through a sequence of *equilibrium* states. *Slow* processes are not necessarily reversible. Bridgman emphasized the intrinsic irreversibility of dislocation motion, the mechanism for irreversible "plastic flow" in metals. Once the barrier to dislocation motion is surmounted, by applying stress, the main part of the stored energy is dissipated in the sudden irreversible relaxation which results.

For fluids, and for solids with the defects frozen in, thermodynamics *is* a useful framework for describing changes of entropy and energy. Further, the usefulness of energy, as a potential for describing the equilibrium of isolated systems, can be extended to the enthalpy, for the equilibrium of adiabatically-isolated systems at fixed *pressure*, and to the two "free energies", Helmholtz' and Gibbs', for describing the equilibria of constant-volume or constant-pressure systems at fixed *temperature*.

More than a century ago, Gibbs followed Boltzmann's lead for gases, relating microscopic mechanics and (q, p) phase-space energy states to macroscopic thermodynamics through statistical mechanics. His linking the two approaches proceeds through a series of three steps: (i) Liouville's incompressible Theorem $(\dot{f}_N \equiv 0)$, is used to motivate the "microcanonical ensemble" description of an isolated Hamiltonian system—a "closed" system in static equilibrium; (ii) this allows dynamical time averages to be replaced by phase-space averages using equal weights for all accessible (q, p) states of the same energy; (iii) *weakly* coupling an ideal gas, for which the phase-space states can be calculated, to a general system, allows both mechanical and thermal equilibrium conditions to be formulated, leading to a new relation for the entropy,

$$S_{\text{Gibbs}} = k \ln \Omega \longrightarrow S = -k\langle \ln f_N \rangle,$$

valid for *all* equilibrium ensembles, not just the microcanonical one. I will consistently refer to this collective N-particle entropy as "Gibbs' ", though the scanty evidence available suggests that both Boltzmann and Gibbs discovered the new relation independently and at about the same time. The evidence—in particular Gibbs' 1884 Philadelphia abstract—is discussed in the biographies of Gibbs written by Klein and Wheeler. Cercignani discusses Boltzmann's 1884 paper. The N-body "Gibbs' entropy" resembles

Boltzmann's one-body dilute-gas entropy, but applies to liquids and solids, as well as to gases, both dense and dilute. It would seem to conflict with Liouville's incompressible phase-flow Theorem, which rules out a changing Gibbs' entropy. This difficulty led Gibbs to discuss irreversibility in terms of a "coarse-grained" approximation to the phase-space density f_N. It is this part of his approach which generated the most criticism. Gibbs' approach made it possible to compute equilibrium properties from phase-space averages, without the need for solving any dynamical equations.

50 years later, Green and Kubo followed Einstein and Onsager, in extending Gibbs' ideas to nonequilibrium systems. They treated linear transport processes in a convincing and complete manner. More recently chaos has provided an answer to those who had questioned the applicability of Gibbs' ensembles to transport theory. Our goal in this Chapter is to summarize statistical mechanics so as to connect thermomechanical microscopic simulations (including nonequilibrium molecular dynamics) with macroscopic descriptions (thermodynamics and continuum mechanics) of material properties.

3.2   Formal Structure of Statistical Mechanics

The fundamental mechanical result which Gibbs used in his equilibrium theory, in order to convert time averages to phase averages, was Liouville's incompressible Theorem. This phase-space flow Theorem is an exact rigorous consequence of Hamilton's equations of motion. Liouville's incompressible theorem makes it possible to introduce probability and use it in the computation of simple equilibrium averages. Liouville's Theorem is the foundation of Gibbs' equilibrium statistical mechanics. The additional tool Gibbs needed was the concept of weak coupling—the idea that two systems could, when so coupled together, access *all* of their states, and in a way which could *violate* Liouville's Theorem, so as to *agree* with the phenomenological predictions of thermodynamics. No doubt Gibbs had in mind a small perturbation, such as a corrugated container, which would introduce what the mathematicians call "mixing", the breaking of correlations between the initial conditions and the current state. In his boxed notes, at the Yale University library, Gibbs suggests putting systems in thermal communication through "slight" gravitational interactions. His concept of "mixing" or "coarse-graining", with the analog of ink in milk, is familiar from his textbook description. We know today that Lyapunov instability provides a natural mechanism for this mixing without the need for any special corrugations, gravitational interactions, or other special mechanical constructions. Gibbs' cautious nature precluded a more complete description of this coupling, but without it, as Krylov was so fond of stating, Gibbs' work was at best incomplete, or at worst wrong.

Consider then, with Gibbs, a two-part system. The larger part is an ideal *gas*, providing a heat reservoir for the smaller part, and coupled to it in such a way as to access all states of both, consistent with a fixed total energy E and total volume V. The ideal-gas reservoir is imagined to be so large that its temperature, T, has negligible fluctuations. The smaller of the two parts making up our combined system could be any small system whatever. For simplicity, we will imagine it to be a *fluid*, with an energy $E_{\text{fluid}} << E$ and a volume $V_{\text{fluid}} << V$. The ideal-gas heat reservoir, with which the fluid interacts, takes up the rest of the energy and volume:

$$E = \mathcal{H}_{\text{gas}} + \mathcal{H}_{\text{fluid}} \; ; \; V = V_{\text{gas}} + V_{\text{fluid}}.$$

Gibbs formulated the distribution of this two-part system among its energy states and was able to derive the properties of a "canonical" (constant temperature T) ensemble directly from this simple two-part "microcanonical" (constant energy) picture. The final result is the equilibrium phase-space probability density:

$$f_{\text{fluid}}(\{q,p\}) \propto e^{[A(N,V,T) - \mathcal{H}(\{q,p\})]/kT}.$$

This "canonical" distribution applies to the smaller of the two linked systems—the fluid in the example just described. It follows from this probability density that the Helmholtz free energy $A \equiv E - TS$ can be determined by working out Gibbs' canonical partition function Z:

$$Z(N,V,T) = e^{-A/kT} = (1/N!)\prod(\int \int dqdp/h)e^{-\mathcal{H}/kT}.$$

The development of this approach was purely theoretical. Neither Gibbs, nor most of his followers, considered the explicit construction of boundary conditions required to contain N particles in a volume V while rendering all states of energy E accessible. They likewise ignored mechanisms to implement the sources and sinks of energy needed to equilibrate systems at a temperature T. Operational computational analogs for these relatively unimportant "formal details" become all-important to any computational application of Gibbs' formulation. Furthermore, the lack of any specific mixing or thermal boundaries provided mathematically-inclined physicists with a quandry—how to be sure that a system described by the Hamiltonian $\mathcal{H}$ would be able to reach all those states included in Gibbs' partition function integral $Z(N, V, T)$.

This quandry launched a variety of investigations into "ergodic theory", a study of necessary and sufficient conditions for phase-space averages and time averages to agree. Sinai obtained a tantalizing result: hard disks and spheres *do* behave in an ergodic mixing manner, even with periodic boundaries. Kolmogorov, Arnold, and Moser obtained a more disquieting (and somewhat more obvious) result: systems described by potentials with a smooth stable bound state cannot generally access all the Gibbs' states at a fixed energy. The lack of any physical boundaries in both these demonstrations is quite unrealistic. Though there is no doubt that complicated boundaries enhance the mixing effects Sinai demonstrated, such boundaries would greatly restrict the applicability of the Kolmogorov-Arnold-Moser Theorem. For a physicist, the clear success of Gibbs' theory suggests that mixing and ergodicity be accepted as basic principles.

Apart from an additive constant E_0, Helmholtz' free energy $A = E - TS$ corresponds to the total energy E of a system—such as our fluid—together with that of a larger heat reservoir, with energy $E_0 - TS$. The temperature T characterizes *both* the system and the reservoir. S is the system entropy. The volume and temperature derivatives of A provide the system's pressure and entropy: $dA = -PdV - SdT$. Gibbs had to show that the Helmholtz free energy following from his canonical partition function integral agreed with experimental pressures and energies. In this he was completely successful. With this achievement, Gibbs' theory could be accepted as exact for the model of Hamiltonian systems. There remained the nagging need to resolve the "ergodicity" quandry, though there was never any real doubt that interesting physical systems are (i) sufficiently perturbed that the initial conditions are unimportant and (ii) sufficiently complicated (with recurrence times exceeding the age of the universe) that the concept of ergodicity is both irrelevant and misleading.

Gibbs' statistical mechanics differed from Boltzmann's kinetic theory by taking place in the full many-body phase space, "gas-space" or "γ-space", while Boltzmann's considerations were mostly restricted to densities in the phase space of a single molecule, "molecule-space" or "μ-space". But, unlike Boltzmann, Gibbs had no results *away* from equilibrium. Let us explore statistical mechanics from the perspective of computer simulations in γ-space. This will allow us to explore the interrelations among thermodynamics, statistical mechanics, and computer simulation. We begin with the initial and boundary conditions.

3.3 Initial Conditions, Boundary Conditions, Ergodicity

Computer simulation is a demanding discipline, in which *all* the "details" included in the computer algorithm must be precisely described. Typical details are (i) the initial conditions, (ii) the boundary conditions, and (iii) the equations of motion. In the atomistic equilibrium systems to which Gibbs' statistical mechanics can be applied, the number of particles, their type, positions, velocities, and the nature of their surroundings, must all be specified. The "surroundings" include (i) a container for the system, with the simplest choice being *periodic* boundaries, as used in an example problem in Section 3.10.3, as well as (ii) any necessary sources and sinks of energy, including forces to perform thermodynamic work and friction coefficients necessary to transfer heat, as described in Section 2.8.

An entire branch of mathematics, ergodic theory, arose in an effort to investigate whether or not a single dynamical trajectory could faithfully represent the states included in Gibbs' microcanonical ensemble. The Ehrenfests believed that it is meaningful to distinguish "quasiergodic" behavior (coming arbitrarily *close* to all states) from "ergodic" behavior (actually *reaching* all states). This distinction is based on the conceptual difficulty inherent in adequately covering a many-dimensional phase-space object with a one-dimensional trajectory. It is one of many examples in which the order of taking limits—small boxes and long times in this case—takes on a significance more apparent than real.

It is evident that ergodic theory has little to do with physics since a small but well-chosen perturbation, a container with a properly-corrugated surface or the gravitational interaction of the system with a small mass outside, would satisfy most physicists' need for an equilibration mechanism. Only in the very simplest of systems, with a few degrees of freedom, is there sufficient time ($\propto e^N$) to access all such states. For example, the time required for an eight-atom system of liquid argon to access all of its quantum states is already comparable to the age of the universe, 10^{17} seconds. Evidently, in view of the impossibility of observing ergodic behavior for most systems, it is actually only important that the fluctuations be sufficiently small. Were this not to be the case, then statistical mechanics would not be a correct description of equilibrium thermodynamics.

Computer simulation of many-body systems began at Los Alamos with the investigation of one-dimensional anharmonic chains of particles. The investigators, Fermi, Pasta, and Ulam, were surprised to find that typical trajectories scrupulously avoided most of the available phase-space energy shell corresponding to Gibbs' microcanonical ensemble. With the benefit of hindsight, we now know that in two- and three-dimensional systems such difficulties are less common, and usually result from an unfortunate combination of force laws and initial conditions. Of course, sufficiently simple systems—a single free particle is the best example—have no hope of passing through (or even near!) *all* the states characterizing a complete Gibbs' distribution *unless* additional thermostat forces are used. An early remedy for the lack of ergodicity was the Langevin equation of Section 2.9, with its irreversible frictional force, $-p/\tau$ and its high-frequency stochastic "noise"—usually Gaussian random forces refreshed at every time step— together providing the desired temperature. Much later, inspired by Nosé's work, Kusnezov, Bulgac, and Bauer* invented relatively simple, robust, and purely-deterministic forces quite capable of imposing Gibbs' canonical distribution on an otherwise free particle.

Excepting such unusual cases, dynamical equilibrium simulations presented no special difficulties. Periodic boundaries could be used to eliminate the unacceptable boundary effects of a physical container. At equilibrium, the statistical properties could be computed by using the "Monte Carlo" sampling technique developed at Los Alamos by Metropolis, the Rosenbluths, and the Tellers. In this way, Gibbs' canonical distribution could be generated by temperature-dependent "moves" of particles. Provided that the initial conditions were wisely chosen, simulations following a few hundred particles for a few tens of thousands of moves accurately reproduced the thermodynamic properties from Gibbs' statistical mechanics.

The good agreement between molecular dynamics and Gibbs' statistics was gratifying. The outcome was not at all obvious until parallel investigations, at Livermore and Los Alamos, applied the two methods to exactly the same systems. Though computers were rapidly gaining speed, many-body simulations have voracious appetites for computer time. In the early 1950s

---

*See their article in the 1990 Annals of Physics for details, examples, and references.

hard elastic disks and spheres were the usual choice for simulation, in order to save computer time and to simplify theoretical interpretations. In the middle 1950s Bill Wood, at Los Alamos, characterized the fluid-solid phase transitions for both hard disks and hard spheres. His Monte Carlo work agreed nicely with simultaneous dynamical studies, at Livermore, carried out by Tom Wainwright and Berni Alder. Wood's mature reminiscences about this work, both scientific and sociological, as given in lectures at two Lake Como summer schools (1985 and 1996), make interesting reading. By 1959 George Vineyard and his colleagues were analyzing fully continuous dynamical models, at Brookhaven, simulating the nonequilibrium relaxation of copper crystals which had been exposed to high-energy radiation.

A generation after Fermi's work on anharmonic chains, dynamic simulations began to be carried out at constant *temperature*, rather than constant energy. The kinetic energy was controlled while the total energy was allowed to fluctuate. This development of isothermal methods was a response to two influences. First, isothermal canonical-ensemble Monte Carlo simulations were commonplace, so that comparison with a dynamical analog was desirable. Second, real laboratory experiments are seldom carried out on isolated systems. Instead they typically use thermal boundaries.

In all such *statistical* canonical ensemble simulations, temperature was defined as usual, by the mean value of the fluctuating kinetic energy: $T \equiv \langle p_x^2/mk \rangle = \langle mv_x^2/k \rangle$. Isokinetic molecular dynamics calculations, with fixed kinetic energy, were the usual dynamical alternative. That approach had been used to model heat reservoirs, first at equilibrium and then in nonequilibrium steady states. The isokinetic dynamics produces an ensemble—different to any of Gibbs'—in which the kinetic energy is fixed while the potential energy is distributed canonically.

Nosé's more-elegant feedback method corresponds *exactly* to Gibbs' canonical distribution. As described in Section 2.8, Nosé's equations of motion, written in the simpler "Nosé-Hoover" form, are:

$$\{\ddot{r} = (F/m) - \zeta\dot{r}\} \; ; \; \dot{\zeta} = [(K/K_0) - 1]/\tau^2.$$

The many-body thermal relaxation time is of order $\sqrt{N}\tau$. This Nosé-Hoover approach is "robust", in that the time-averaged kinetic temperature is maintained *exactly*, even in the presence of additional external driving forces which can convert the *smooth canonical distribution* to a *complex multifractal attractor*.

Systems, such as the harmonic oscillator, which are even farther from ergodic than Fermi's nonlinear chains, require additional control variables to reach the complete canonical distribution. A simple workable approach is to use *two* control variables $\{\zeta, \xi\}$, rather than just one. Consider a simple one-dimensional harmonic oscillator. For convenience choose the mass, force constant, temperature, and characteristic thermostat times all equal to unity. Two alternative examples of this generalized approach to the oscillator's thermomechanics are the following:

$$\{\dot{q} = p \; ; \; \dot{p} = -q - \zeta p - \xi p^3 \; ; \; \dot{\zeta} = p^2 - 1 \; ; \; \dot{\xi} = p^4 - 3p^2\},$$

$$\{\dot{q} = p \; ; \; \dot{p} = -q - \zeta p \; ; \; \dot{\zeta} = p^2 - 1 - \xi\zeta \; ; \; \dot{\xi} = \zeta^2 - 1\}.$$

Numerical investigations—for which see my papers with Holian and Posch, 1996 and 1997, respectively—have established that *both* these approaches provide the complete canonical distribution for the harmonic oscillator energy, $(q^2 + p^2)/2$, with independent Gaussian distributions for the two thermostat variables $\{\zeta, \xi\}$:

$$f \propto e^{-(q^2+p^2)/2} e^{-(\zeta^2+\xi^2)/2}.$$

Generalizing Nosé's approach to *quantum* systems remains a pressing need. A variety of approaches has been tried. Some approaches use feedback, adjusting the wavefunction or its gradient in such a way as to promote thermal equality between system and thermometer. Some use "Gaussian random matrices" to represent thermostats. See the articles by Jürgen Schnack and by Dimitri Kusnezov for contemporary accounts.

3.4 From Hamiltonian Dynamics to Gibbs' Probability

In phase space, the coordinates of a single multidimensional point give the complete microscopic description $(\{q\}, \{p\}, \{\zeta\}, \dots)$. For *Hamiltonian* systems without heat sources and sinks there is no expansion or contraction of the phase-space flow. The comoving probability density is incompressible. And, according to Gibbs' equilibrium statistical mechanics, the logarithm of this density, when properly averaged, gives the entropy. Without any further assumptions or considerations, it would appear therefore that Hamilton's mechanics, though fine *at* equilibrium, provides no *approach* to it. Gibbs' entropy cannot change.

It is Liouville's *incompressible* theorem which shows that both the co-moving phase volume $\otimes$—the "extension in phase"—and the comoving phase-space probability density f flow through phase space unchanged. In the usual *Cartesian* case, all the components of the phase-space velocity divergence vanish:

$$\{\dot{x} = (p_x/m) \longrightarrow (\partial\dot{x}/\partial x) = 0 \; ; \; \dot{p}_x = F(\{x\}) \longrightarrow (\partial\dot{p}_x/\partial p_x) = 0\}.$$

Even with arbitrary *generalized* coordinates the *summed-up* terms vanish:

$$\{(\partial\dot{q}/\partial q) + (\partial\dot{p}/\partial p) = (\partial/\partial q)(+\partial\mathcal{H}/\partial p) + (\partial/\partial p)(-\partial\mathcal{H}/\partial q) \equiv 0\}.$$

For either type of coordinates the generalized velocity of the flow through phase space, $v \equiv (\dot{q}, \dot{p})$ has no divergence. Thus *any* comoving element of phase-space volume $\otimes \equiv \prod(dqdp)$, small or large, is conserved by the flow, and so must be the corresponding probability density $f(\{q, p, t\})$. Either of these two results implies the other because the product $f\otimes$ corresponds precisely to probability, and so must be conserved by *any* flow equations. Both components, f and $\otimes$ separately, are unchanged in Hamiltonian flows.

Complete thermodynamic equilibrium has three components, mechanical, thermal, and chemical. Equilibrium is characterized by constancy of pressure, temperature, and all the chemical potentials. No net accelerations, heat currents, or chemical reactions occur at equilibrium. This static equilibrium situation corresponds, in Gibbs' view, to a dynamic ensemble in an unchanging stationary state of flow, with a fixed probability density $f(q, p, t)_{eq} \to f(q, p)$ everywhere in phase space: $\partial f/\partial t \equiv 0$. In an ergodic isoenergetic *mixing* system, the conclusion that the probability density f does not change with time suggests that it can only be constant, in the entire accessible region. Otherwise, the density at a fixed phase-space point would change with time, and could not characterize equilibrium. A thin energy shell of the constant probability density which results defines Gibbs' "microcanonical ensemble". This "ensemble" can be viewed as a collection of systems distributed according to the constant invariant probability density $f_{eq}(q, p, t) = f(E)$. In order for Gibbs' ensemble to represent the long-time-averaged properties of an arbitrarily selected dynamical system—what Gibbs called the "time ensemble"—it is necessary that the dynamics be *mixing*, so that the correlation of long-time-averaged properties with initial conditions is eventually lost.

3.5 From Gibbs' Probability to Thermodynamics

Before Alder, Wainwright, and Wood's computer simulations of the 1950s it was entirely unknown whether or not Gibbs' static statistical ensemble averaging would agree with long-time dynamical averages based on Newton's equations of motion. Until then intuition was crippled by a lack of examples. Arguments raged over the existence and significance of "holes" and "cages" in liquids. Even the existence of a solid phase for hard spheres was controversial. The computer simulations dissipated this foggy atmosphere. Alder, Wainwright, and Wood found that time averages, for fluid and solid systems of a few hundred hard disks or spheres, were in excellent agreement with Gibbs' statistical phase-space theory, even including the "number-dependent" effects which cause small systems to deviate (relatively slightly) from large-system "thermodynamic behavior".

In Gibbs' theory entropy is of primary significance. Gibbs had established the identity of thermodynamic entropy with the microcanonical phase volume, $S = k \ln \Omega(N, E, V) = -k \langle \ln f_N \rangle$. He also generalized this microcanonical relation to other equilibrium ensembles with different phase-space probability densities: $S_{\text{Gibbs}} = -k \langle \ln f \rangle$. His demonstration that thermodynamic entropy corresponds *generally* to phase-space probability density is a two-step process. First, the relation is established in the microcanonical constant-energy case. Second, the same relation is shown to hold in other ensembles. Let us follow Gibbs' argument in the microcanonical case. We begin by considering the interaction of an ideal gas with another system. The ideal gas is chosen because its phase volume can be calculated analytically:

$$\Omega_{\text{gas}} \equiv (N!)^{-1} \prod (\int \int dq \, dp / h) \propto (V/N)^N (E/N)^{ND/2}.$$

The First Law, $\Delta E = \Delta Q - \Delta W$, combined with the ideal-gas equation of state gives the ideal-gas pressure and temperature in terms of the volume and energy derivatives of Ω_{gas} :

$$PV = (DE/2) = NkT \longrightarrow N/V = P/kT = (\partial \ln \Omega / \partial V)_E;$$

$$\Delta E = T\Delta S - P\Delta V \longrightarrow ND/2E = 1/kT = (\partial \ln \Omega / \partial E)_V.$$

I omit subscripts on Ω here because the last relations, which link the pressure and temperature to the phase volume Ω_{gas}, actually hold for *any* fluid. To see this, imagine coupling the ideal gas, weakly, to an arbitrary equilibrium fluid, in such a way that the total volume and energy are fixed.

$$V_{\text{total}} \equiv V_{\text{gas}} + V_{\text{fluid}} \; ; \; E_{\text{total}} \equiv E_{\text{gas}} + E_{\text{fluid}}.$$

Assuming ergodic mixing, the states available to such a combined, but weakly-coupled, system are given by the product, $\Omega_{\text{gas}}\Omega_{\text{fluid}}$. Thus:

$$\ln \Omega_{\text{total}} = \ln \Omega(E,V)_{\text{gas}} + \ln \Omega(E,V)_{\text{fluid}}.$$

It is easy to see that *maximizing* $\ln \Omega_{\text{total}}$ (to find the equilibrium conditions relating $\ln \Omega_{\text{gas}}$ to $\ln \Omega_{\text{fluid}}$), gives *equal* derivatives *at* the maximum:

$$(\partial \ln \Omega_{\text{gas}}/\partial V)_E = (\partial \ln \Omega_{\text{fluid}}/\partial V)_E = P_{\text{eq}}/kT_{\text{eq}} \; ;$$

$$(\partial \ln \Omega_{\text{gas}}/\partial E)_V = (\partial \ln \Omega_{\text{fluid}}/\partial E)_V = 1/kT_{\text{eq}}.$$

These two conditions on the partitioning of the volume and energy between gas and fluid correspond to the thermodynamic conditions of mechanical and thermal equilibrium. Their identification from the maximum condition $\delta \ln \Omega = 0$ requires the neglect (abundantly justifiable through the Central Limit Theorem as is detailed in Section 3.8) of fluctuations in the neighborhood of the maximum.

Because $k \ln \Omega$ is a state function, depending upon E and V and with the *same derivatives* with respect to these variables as the thermodynamic entropy, it is clear that, apart from a conventional additive constant[†] the entropy is *exactly* $k \ln \Omega$. $k \ln \Omega$ has exactly the same properties as the thermodynamic entropy S: (i) it is a state function, (ii) it is additive, for weak, but mixing, coupling, (iii) it is a maximum at equilibrium, (iv) its isoenergetic volume derivative is P/T, and (v) its isochoric energy derivative is $1/T$. The correspondence between the microscopic and macroscopic formulations of entropy is complete. The only lingering question, to which we will return repeatedly, is the compatibility of Liouville's Theorem with the description of *irreversible* processes.

The *canonical* ensemble, with an isothermal phase-space probability density $f \propto e^{[A-\mathcal{H}]/kT}$, gives an alternative, and computationally more useful, relation between dynamics and thermodynamics. It comes from the imposition of "weak coupling" linking a large thermostat, or heat reservoir,

---

[†]Quantum mechanics provides the constant through the *Third* Law of Thermodynamics.

to a relatively smaller fluid subsystem with Hamiltonian $\mathcal{H}_{fluid}$. The thermostat and fluid together make up a two-part system described by Gibbs' microcanonical ensemble.

3.6 Pressure and Energy from Gibbs' Canonical Ensemble

From the microscopic point of view, the energy is not at all mysterious. It is natural to write it, as we have done continually, as the sum of kinetic and potential parts, $\mathcal{H} = K + \Phi$. Here we consider the simplest case, in which the N-body potential energy is a sum of pair terms,

$$\Phi(\{r_i\}) = \sum_{i<j} \phi(\{r_i - r_j\}).$$

For simplicity in what follows, I leave off the explicit dependence of the summed pair potentials $\sum \phi = \Phi$ on the separations of all pairs of interacting particles $\{i < j\}$. Gibbs' canonical partition function,

$$Z(N, V, T) \propto \int \int e^{-\mathcal{H}/kT} dq^{3N} dp^{3N},$$

can be explicitly differentiated with respect to volume and temperature. The resulting expressions match the derivatives of the macroscopic thermodynamic *Helmholtz'* free energy,

$$d(A/kT) = -(P/kT)dV - (E/kT^2)dT \longrightarrow dA(N, V, T) = -d(kT \ln Z).$$

Apart from the conventional additive constant the microscopic analog of Helmholtz' free energy is the logarithm of the microscopic canonical partition function Z, multiplied by $-kT$.

The temperature derivative of $\ln Z$ gives the expected expression for the thermodynamic energy:

$$E = kT(\partial \ln Z/\partial \ln T)_V = \langle \Phi \rangle + \langle K \rangle,$$

where the averages are carried out with a probability density proportional to $e^{-\mathcal{H}/kT} = e^{-\Phi/kT} e^{-K/kT}$. The volume derivative gives the "virial theorem" expression for the hydrostatic pressure in a fluid with D space dimensions:

$$PV = (1/D)\sum_N [\langle -r \cdot \nabla \Phi \rangle + \langle p \cdot p/m \rangle].$$

The tensor form, with the centered dots, angular brackets, and D removed, follows much more directly from a purely *mechanical* evaluation of the instantaneous momentum flux[‡]. The latter pressure expression, and its analog for the heat flux, are invaluable in microscopic simulations.

3.7 Gibbs' Entropy *versus* Boltzmann's Entropy

Until Section 3 of his 1884 paper, "Über die Eigenschaften monozyklischer und anderer damit verwandter Systeme", Boltzmann's gas-phase discussions of entropy all took place in single-molecule phase space, "μ space" $\{x, y, z, p_x, p_y, p_z\}$. In μ space Boltzmann considered cells large enough to contain several molecules—so that Stirling's approximation could be used for the corresponding factorials, $\{N_{cell}!\}$. The μ-space "Boltzmann entropy" is $-Nk\langle \ln f_1 \rangle$, where f_1 is the *one*-particle probability density. In his 1884 paper Boltzmann formulated both "Gibbs' " microcanonical and canonical ensembles, described in the full *many*-body phase space, "γ space", rather than the single-molecule μ space. The many-body formulation leads naturally to "Gibbs' entropy", $-k\langle \ln f_N \rangle$.

It appears that the many-body Gibbs' entropy was actually the independent discovery of both men, though their disparate styles make it hard to be sure. Boltzmann left us a voluminous record documenting his changing views of kinetic theory. Gibbs left us very little trace of his own evolving ideas. We have only his Philadelphia lecture abstract from 1884, his book from 1902, and a handful of notes, mostly without dates, at the Yale University library. Gibbs certainly delivered a lecture at the American Association for the Advancement of Science meeting in Philadelphia in 1884: "On the Fundamental Formula of Statistical Mechanics with Applications to Astronomy and Thermodynamics". According to Klein this is the first appearance of the words "Statistical Mechanics" in print. The published abstract stresses the importance of the incompressible form of Liouville's Theorem, discussed in Section 2.5. Neither in this abstract nor in his book, which appeared eighteen years later, does Gibbs refer to Liouville. But it is quite clear that he had made the connection between the many-body phase-space density and entropy by 1884. He then worked out a detailed exposition over the next 18 years.

Gibbs' reticent style is nicely matched by Onsager's, as a visit to their graves near Yale reveals. Even though Gibbs sent nearly all of his published

---

[‡]See Section 5.5 of my *Computational Statistical Mechanics* for a detailed argument.

work to Boltzmann I could find no record of their correspondence beyond a rather formal invitation from Boltzmann, which Gibbs declined, to attend an 1892 meeting in Nuremburg.

Distributions in the six-dimensional single-molecule μ space are evidently much simpler than those in the $6N$-dimensional many-body γ space. It is evident that the many-body phase volume available to the strongly-interacting molecules in a condensed phase is greatly reduced by their short-ranged repulsive interactions. We can estimate this effect quantitatively by using the Mayers' theory to analyze van der Waals' picture of hard spheres. Because the centers of no two spheres can come more closely together than the sphere diameter σ, there is a minimum "closest-packing" volume per sphere, $\sigma^3/\sqrt{2}$, at which the many-body partition function must vanish. The Mayers' virial expansion of the canonical partition function makes it possible to estimate the large-N limit of the phase volume for a dense fluid of N such spheres, at a temperature T and in a volume V:

$$e^{+\langle \mathcal{H}/kT \rangle} \int dq^{3N} \int dp^{3N} e^{-\mathcal{H}/kT} \simeq$$

$$V^N[1 - b\rho + 0.1875(b\rho)^2 + 0.0502(b\rho)^3 + \ ... \ ]^N (2\pi e m k T)^{3N/2}/N! \ ;$$

$$b\rho = 2\pi N\sigma^3/3V \ ; \ e \equiv 2.718281828.$$

Though it is evident that Boltzmann correctly *formulated* N-body thermodynamic properties in his 1884 paper, it must be emphasized that his H-Theorem explanation of irreversible processes is restricted to low-density gases. The H-Theorem is an approximate demonstration—for isolated gases—that the one-body "Boltzmann entropy",

$$S_B(t) \equiv -Nk\langle \ln f_1(r, v, t) \rangle,$$

cannot decrease with time. A convincing mechanical proof of the Second Law of Thermodynamics for condensed matter awaited the advances of Green and Kubo.

To illustrate the important distinction between Gibbs' and Boltzmann's entropies, Prigogine, Kestemont, and Mareschal simulated an isolated system in which Boltzmann's entropy decreases while Gibbs' does not[§]. They began with an expanded two-dimensional crystal of particles with a short-ranged *purely-repulsive* interaction $\phi(r < \sigma)$. They argued as follows:

[§]This work appears in the 1989 *Complex Flows* volume edited by Michel Mareschal.

choose the nearest-neighbor spacing just *exceeding* the range of the potential σ, corresponding to the density ρ and select the initial particle velocities from the Maxwell-Boltzmann equilibrium distribution for the temperature T: $e^{-mv^2/2kT}$. The potential energy vanishes so that the Boltzmann entropy corresponds to that of an ideal gas, $S_{\text{ideal}}(\rho, T)$. The *motion* of the system, at fixed density and energy E, soon leads to a state with a *positive* potential energy $\langle \Phi \rangle \equiv \langle \sum \phi \rangle > 0$ and a correspondingly reduced kinetic energy, $K = E - \Phi$. The Boltzmann entropy then suffers a reduction, while Gibbs' entropy is unchanged.

An equally-dramatic example of apparent Boltzmann-entropy loss occurs during an *inelastic* collision of two or more solid metallic spheres. Two such spheres, composed of a few hundred metal atoms each, are evidently able to access their entire container volume V independently, *so long as they do not interact*. Numerical simulations, for a two-dimensional analog of this situation—the collision of four similar metallic bodies—can result in an inextricably cold-welded composite particle. See Figure 3.1.

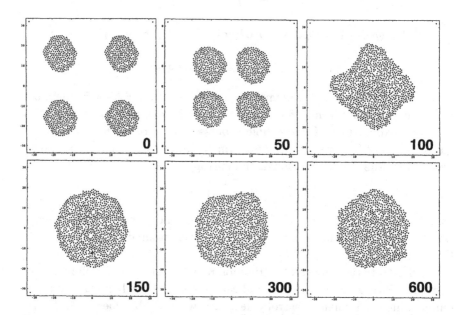

Figure 3.1: Snapshots from a simulation illustrating the cold welding of four colliding two-dimensional metallic drops. In units of the final-state sound traversal time the six snapshots correspond to times of $\{0, 1, 2, 3, 6, 12\}$.

3.8 Number-Dependence and Thermodynamic Fluctuations

A real physical boundary necessarily introduces a host of surface contributions to the energy and the other extensive properties of an enclosed system. These surface contributions are of the order of the surface area $\propto N^{(D-1)/D}$ in D-dimensional systems. They can neatly be avoided by the use of *periodic* boundaries, for which the number-dependent contributions are considerably smaller, typically of order unity. This small systematic number-dependence for periodic systems can be thoroughly and systematically understood for simple models. This is the case for hard spheres, for which the Mayers' cluster-integral expansion has been evaluated through the first seven terms. See the example given in Section 3.10.2.

Gibbs' statistical mechanics makes it relatively easy to calculate equilibrium fluctuations. The energy and enthalpy fluctuations, for example, are proportional to the isochoric and isobaric heat capacities:

$$\langle \delta E^2 \rangle = \langle (E - \langle E \rangle)^2 \rangle = C_V k T^2 \; ; \; \langle \delta H^2 \rangle = \langle (H - \langle H \rangle)^2 \rangle = C_P k T^2.$$

These examples illustrate the general rule that fluctuations of intensive quantities become negligible in sufficiently-large classical systems:

$$\langle (\delta E/N)^2 \rangle \simeq \langle (\delta H/N)^2 \rangle \propto (kT)^2/N.$$

Sufficiently-large means that the parts can be treated as independent, so that the instantaneous state of a simple large system corresponds to many—enough for the Central Limit Theorem to apply—independent small-system samples. The Central Limit Theorem guarantees that large-system equilibrium fluctuations have Gaussian distributions, and become negligibly small in amplitude, $\propto \sqrt{1/N}$, as the system size increases.

3.9 Green and Kubo's Linear-Response Theory

Linear-Response Theory is based on Gibbs' statistical mechanics. It provides a partial explanation of irreversible behavior. His "canonical distribution":

$$f_{NVT}(q,p) \propto e^{-\mathcal{H}/kT},$$

makes it possible to treat ensemble-averaged flows of mass, momentum, and energy as small perturbations. From this point of view equilibrium, represented by a Gibbs ensemble, is characterized by opposing currents, which largely cancel. The residual currents in individual systems, averaged over a large volume with N particles, are of order $\sqrt{1/N}$. The most likely situation, corresponding to the ensemble average, is that the positive and negative contributions to the mass current, $\pm\langle\rho(|v_x|,|v_y|,|v_z|)/2\rangle$, precisely cancel. This delicate balance can be offset by any perturbation (such as a gradient in concentration, velocity, or temperature), giving rise to a *net* current proportional to the perturbation (the "linear" response).

The simplest illustration of the Green-Kubo theory is a shear flow in the xy plane, with the macroscopic laboratory-frame flow velocity in the x direction proportional to the y coordinate:

$$\langle \dot{x} + (p_x/m)\rangle = \dot{\epsilon}y.$$

Formally, this flow can be generated by using a perturbed "Doll's-Tensor" Hamiltonian which includes the strain rate $\dot{\epsilon}$:

$$\mathcal{H}_{\text{shear}}(q,p,\dot{\epsilon}) = \mathcal{H}_{\text{equilibrium}}(q,p) + \dot{\epsilon}\sum yp_x.$$

Periodic boundary conditions can be constructed in the usual way, provided that the adjacent images in the y direction are displaced horizontally:

$$\{x(y \pm L, t)\} = \{x(y,t) \pm \dot{\epsilon}Lt\}.$$

This periodic shear-flow example is worked out in Section 8.6. Green and Kubo's theory[¶] shows that the limiting small-strain-rate shear viscosity in this example is given by the *equilibrium* integral of the shear stress auto-correlation function:

$$\eta = (V/kT)\int_0^\infty \langle P_{xy}(0)P_{xy}(t)\rangle_{\text{eq}}dt \equiv (V/kT)\int_0^\infty \langle \sigma_{xy}(0)\sigma_{xy}(t)\rangle_{\text{eq}}dt.$$

Figure 3.2 shows the correlation function for a typical dense fluid. The product $\langle P_{xy}(0)P_{xy}(t)\rangle$ is of order $(1/N)$.

[¶]Helfand's 1960 Physical Review paper provides a very nice approach to this problem.

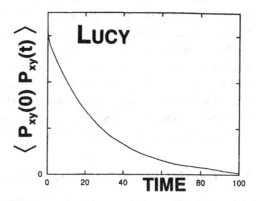

Figure 3.2: Green-Kubo equilibrium shear stress correlation function. The integral gives the shear viscosity coefficient, η.

The details of linear-response theory, formulated by Green and Kubo and Onsager, are in agreement with Boltzmann's kinetic theory for gases, but are applicable also to dense fluids and solids as well as to quantum systems. The simple gas-phase case provides a qualitative explanation of irreversible behavior: when the local equilibrium distribution varies, due to the gradients $\nabla\rho, \nabla v, \nabla T$, the resulting mass, momentum, and energy flows produce perturbations proportional to the products of these gradients and an effective distance between collisions (or "mean free path"). The time required for these perturbations to be established is the time between collisions, while the time required for complete equilibration is much longer, $\propto L^2/D$ for a system of size L with a diffusive transport coefficient $D[\text{meters}^2/\text{second}]$. Although these arguments are not at all rigorous, extensive comparisons of the results of linear-response theory with experiments and with direct nonequilibrium computer simulations confirm their validity.

3.10 Example Problems

The first two example systems described here are prototypes for the discussion of the equilibrium properties of solids and fluids. The "quasiharmonic" approximation to the properties of a crystalline solid, or a glass, replaces the potential part of the Hamiltonian by a (positive) quadratic form describing the interactions of particles through small-displacement force constants. Because such a positive quadratic Hamiltonian can be routinely diagonalized, to give a sum of "normal-mode" Hamiltonians, with a simple product partition function, $Z_{NVT} = \prod z_{\text{mode}}$, this approach to Gibbs' canonical partition function is straightforward.

Hard spheres, particles which cannot overlap, likewise provide a relatively simple approximation to Gibbs' partition function for dense fluids. This approximation proceeds by a systematic evaluation of the Mayers' two-body, three-body, ... corrections to the single-particle ideal-gas partition function. The third and final problem in this Section demonstrates the need for coarse-graining in understanding a classic irreversible flow process, the free expansion of a pressurized fluid into a larger container.

3.10.1   *Quasiharmonic Thermodynamics*

A static zero-temperature solid can be confined, with periodic boundary conditions, at a pressure given by the derivative of the potential energy with respect to volume,

$$P_0 \equiv -d\Phi_0/dV.$$

The thermal energy required to heat such a cold solid to a temperature T is of order NkT. The effect of this additional energy on the pressure can conveniently be described by the phenomenological Grüneisen equation of state:

$$P(T) - P_0 \equiv \gamma_{\text{Grüneisen}}(E - \Phi_0)/V.$$

Such a picture develops naturally if the energy is considered to be a sum of the minimum possible pair-term energy plus additional thermal contributions to the kinetic and potential energy, $DNkT/2$ each in D dimensions.

A classical crystal is completely motionless at zero temperature, with neighboring particles' relative motion always small compared to the interparticle spacing. Under these conditions a useful approximation for the energy is the "cold" curve, $E(T = 0, V) = \Phi_0(V)$, with the *thermal* part of the energy added on. If the thermal motions are relatively small, the dependence of the full N-body potential energy on particle displacements away from the minimum-energy structure, $\{\delta\}$, can be truncated after the quadratic terms:

$$\Phi(\{r\}) = \Phi_0 + (1/2)\sum \delta_i \cdot \kappa_{ij} \cdot \delta_j.$$

Evidently the force constants $\{\kappa\}$ in this truncated potential give rise to forces which are *linear* in all the particle displacements, so that the whole problem reduces to a linear one, which can be solved by superposing N independent solutions. The motion governed by such a potential is a linear combination of normal-mode vibrations. The motion is conventionally termed "quasiharmonic", rather than "harmonic", so as to emphasize the fact that both the static energy Φ_0 *and* the set of force constants $\{\kappa_{ij}\}$ depend upon the volume at which the expansion is carried out. We illustrate the power of Gibbs' statistical mechanics by considering the simplest interesting example of this problem, the thermodynamics of a two-dimensional harmonic crystal at finite temperature.

Imagine a two-dimensional triangular lattice with periodic boundary conditions. For simplicity suppose that only nearest neighbors interact. The vibration of a single particle in this lattice, with all of its neighbors fixed at their lattice sites, defines the one-particle "Einstein frequency" ω_{Einstein}. If the nearest-neighbor interaction is a purely-harmonic Hooke's-law interaction, with a fixed constant force constant κ, the Einstein frequency is proportional to $\sqrt{\kappa/m}$:

$$\omega_{\text{Einstein}} = 2\pi\nu_{\text{Einstein}} \equiv \sqrt{3\kappa/m}.$$

For the two-dimensional system the "Einstein approximation" to Gibbs' canonical partition function is:

$$Z_{\text{Einstein}} = e^{-\Phi_0/kT} \prod_{2N} z_{\text{Einstein}} = e^{-\Phi_0/kT}(kT/h\nu_{\text{Einstein}})^{2N}.$$

An *exact* solution of the small-vibration problem remains relatively simple when *all* the particles move simultaneously, provided that periodic boundaries are used. The zero-pressure solution (adjacent particles are separated, on the average, by the equilibrium separation, where the first derivative of the pair potential vanishes) was first worked out numerically, by evaluating the normal-mode frequency distributions of finite periodic N-particle crystals with $N = 2n^2$ for $2 \leq n \leq 20$. This numerical work made it possible to estimate the large-N "thermodynamic limit" corresponding to an *infinite* number of particles with five-digit accuracy. Dale Huckaby soon carried out the same calculation analytically[||]. The perfect agreement between the numerical calculation and Huckaby's analysis provided evidence that the usual practice of extrapolating small-system computer results to

[||] For references see my paper in the 1 September 1972 Journal of Chemical Physics.

the large-system limit is valid. The result of this work showed also that the cooperative motion of the particles increased Gibbs' canonical partition function by a numerical factor of $e^{0.273N}$ relative to the naïve prediction of the Einstein model:

$$Z_{\text{exact}}^{N \to \infty} = e^{0.273N} e^{-\Phi_0/kT} (kT/h\nu_{\text{Einstein}})^{2N}.$$

The exact oscillation frequencies *are* typically volume-dependent, justifying the Grüneisen prescription for the pressure, which includes contributions proportional to $\langle d\ln\nu/d\ln V \rangle$. Figure 3.3 shows the rather intricate form of the exact frequency distribution.

Figure 3.3: Vibration frequency distribution for a two-dimensional triangular lattice with harmonic first-neighbor interactions.

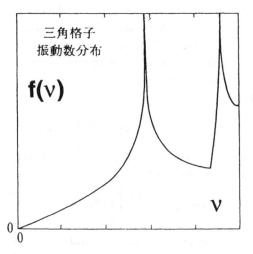

3.10.2   *Hard-Sphere Thermodynamics*

Hard particles are relatively easy to treat using Gibbs' statistical mechanics. In one dimension the exact partition function can be calculated, just as it can be for a harmonic chain. In two and three dimensions no complete partition functions are available for more than seven particles. But corresponding series expansions *can* be obtained from the Mayers' virial theory. The results for hard disks and spheres in the fluid phase are relatively simple. The Mayers' theory, based on the ideal gas and the perturbation function:

$$f_{\text{Mayers}} \equiv e^{-\phi(r)/kT} - 1,$$

provides a formal density-series expansion for all the thermodynamic properties. For hard spheres the first seven terms are known[**]. The "compressibility factor" PV/NkT depends only on density:

$$PV/NkT = 1 + b\rho + 0.625(b\rho)^2 + 0.287(b\rho)^3 + 0.110(b\rho)^4 + 0.037(b\rho)^5 + \ldots .$$

Figure 3.4 shows the convergence of this series expansion to the results from computer simulations (shown as open circles). It is to be emphasized that the alternative *direct* measurement of the time-averaged pressure, using molecular dynamics to evaluate the tensor form of the virial theorem:

$$PV = \Big\langle \sum_{i<j}(rF)_{ij} \Big\rangle + \Big\langle \sum_{i}(pp/m)_i \Big\rangle,$$

is actually much more efficient, and accurate, than is the numerical evaluation of the Mayers' series expansion.

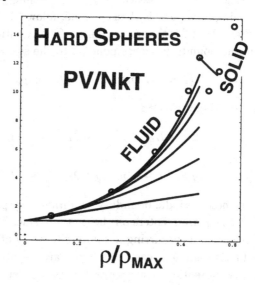

Figure 3.4: Convergence of density series for PV/NkT. The higher-density solid phase is stable at densities above 2/3 the close-packed. The fluid curves indicate sums including from one to seven coefficients.

The hard-sphere "solid" equation of state is shown too, for comparison. The theory for this phase is not so well developed as the Mayers'—the problem is that the quasiharmonic expansion of the previous example does not apply to the singular hard-sphere interaction. Francis Ree and I worked out the properties of the hard-sphere fluid and solid phases in 1968, so as to

[**]For references, and a start on the *eighth* term see the paper by Foidl and Kasperkowitz.

locate the transition linking them more precisely. The solid-phase pressure was also evaluated from the virial theorem, applied to a system with an imposed face-centered-cubic solid structure. See again Figure 3.4 as well as our 1968 paper.

3.10.3 *Time-Reversible Confined Free Expansion*

With modern work stations many-body molecular dynamics simulations with up to 50,000 particles are today fully as feasible as were few-body simulations in the 1950s. With parallel computers simulations can treat tens or hundreds of millions of particles. To illustrate a many-body molecular dynamics simulation, let us consider the prototypical macroscopic irreversible process, the confined free expansion of a pressurized many-body system. We begin with a classical dense fluid confined to an $L \times L$ square. The confining boundary is then released, allowing the particles to move within a larger $2L \times 2L$ square container. The equations of motion are simple, if periodic boundaries corresponding to the square container are used:

$$\{\dot{x} = (p_x/m) \; ; \; \dot{y} = (p_y/m) \; ; \; \dot{p}_x = F_x : \dot{p}_y = F_y\},$$

with continual checks:

$$\{x < -L \longrightarrow x = x + 2L \; ; \; x > +L \longrightarrow x = x - 2L\};$$

$$\{y < -L \longrightarrow y = y + 2L \; ; \; y > +L \longrightarrow y = y - 2L\}.$$

With these periodic boundary conditions any particle leaving the confining $2L \times 2L$ square reënters on the opposite side. Equivalently, the motion can be regarded as occurring in an infinite array of identical $2L \times 2L$ squares.

In the simplest case the forces are negative gradients of a pairwise-additive potential. The Figure shows several snapshots in the time evolution of 16,384 particles interacting with Lucy's short-ranged purely-repulsive pair potential:

$$\phi_{\text{Lucy}}(r < 6) = (5/36\pi)[1 + (r/2)][1 - (r/6)]^3.$$

This potential function reduces numerical integration errors because it has two continuous derivatives at the cutoff distance, $r = 6$. Because this potential is also a smooth-particle "weight function", as described in Section

6.6, it can be used to compute interpolated variables throughout space.

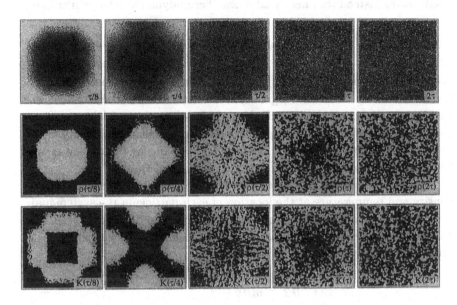

Figure 3.5: Snapshots showing the confined free expansion of 16,384 particles appear in the top row. Corresponding density and kinetic-energy contours, with black indicating below-average values and white above-average, are shown below. The maximum time shown is two sound traversal times. For details see the paper by Harald Posch and me, in the February 1999 issue of Physical Review E **59**, pages 1770-1776.

The many-body particle motions can be made fully reversible by using a *bit-reversible* algorithm to advance the coordinates [the simplest example is Levesque and Verlet's algorithm described in Section 2.3]:

$$\{r_+ = 2r_0 - r_- + [F_0(\Delta t^2/m)]_{\text{integer}}\}.$$

A naïve hydrodynamic analysis of this problem suggests that either viscosity or heat conductivity could dissipate the expansion energy into heat in a time of order L^2/D, where D is the kinematic viscosity (η/ρ) or thermal diffusivity $(\kappa m/\rho C)$. Both diffusivities have units [meters^2/second]. If the entropy of the equilibrating system is estimated by summing up local entropies based on local temperatures (velocity fluctuations) and densities the results show instead that nearly all of the thermodynamic entropy increase,

$$\Delta S = Nk \ln(V_{\text{final}}/V_{\text{initial}}) = Nk \ln(4L^2/L^2) = Nk \ln 4,$$

occurs in a *much* shorter time, less than a single sound-traversal time. Evidently linear hydrodynamics greatly overestimates the decay time. See Figure 3.6. Although the periodic boundaries allow the particles to interpenetrate, simulations of the equilibration of steep sinusoidal density distributions can lead to very similar results, and without any interpenetration. See the 1999 paper by Hoover, Posch, Castillo, and Hoover.

The entropy calculations were carried out by evaluating a coarse-grained density and temperature at the location of each particle in the system and summing the corresponding local-equilibrium entropy densities, $\{s(\rho, T)\}$. Unless the *local* velocity fluctuations, in the neighborhood of each particle, are correctly taken into account, with

$$2kT \equiv m\langle(v - \langle v \rangle)^2\rangle,$$

there is no entropy increase at all. Accounting for the fluctuations—missing in the usual hydrodynamic treatment—by measuring both $\langle v \rangle$ and $\langle v^2 \rangle$ separately, the entropy increase agrees well with that calculated using Gibbs' statistical mechanics, $Nk \ln 4$ for N particles.

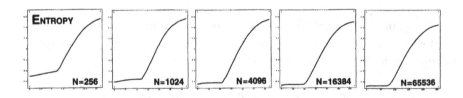

Figure 3.6: Increase of entropy with time for confined free expansions using from 256 to 65,536 particles. The abscissæ correspond to one half sound traversal time. The ordinates correspond to entropy increases of $Nk \ln 4$.

3.11 Summary

Gibbs' statistical mechanics provides a general phase-space representation for the equilibrium thermodynamic entropy:

$$S = S_{\text{Gibbs}} \equiv -k\langle \ln f_N \rangle.$$

It appears that Boltzmann, who invented the one-body "Boltzmann entropy", $S_{\text{Boltzmann}} \equiv -Nk\langle \ln f_1 \rangle$ much earlier, arrived at an *identical* formulation of the N-particle "Gibbs'" entropy *independently*, and at about the same time as Gibbs. From the N-body entropy, together with the thermodynamic statement of the First and Second Laws for *reversible* processes, $TdS = dE + PdV$, all of the other thermodynamic quantities can be obtained. Computer simulations—first for hard spheres, but later for a host of different materials—have demonstrated the simplicity and usefulness of this approach. Gibbs' ensemble theory also provides an instructive formulation of fluctuations about the most-likely state, showing that the thermodynamic effects of fluctuations vanish as system size increases, in accord with the Central Limit Theorem.

Gibbs' approach made possible Green and Kubo's exact formulation of the linear response of equilibrium systems to nonequilibrium perturbations. This formulation provides a basis for the algorithms used in *nonequilibrium* simulations, together with an understanding of the link between reversible dynamics and ensemble-averaged dissipative processes. Although Gibbs was careful to avoid a precise description of the coupling required to achieve thermal equilibrium, time-reversible thermostats, providing deterministic coupling, came into common use with the development of fast computers.

Chapter 4

Irreversibility in Real Life

Oh come with old Khayyám, and leave the wise
to talk; one thing is certain, that life flies;
One thing is certain, and the rest is lies—
the flower that once has blown forever dies.

Omar Khayyám, as translated by Edward FitzGerald

4.1 Introduction

"Irreversible" means "impossible to annul or to run backward". And life,
physics, and nature, are all like that. Velocity and temperature differences
tend to decay whenever they substantially exceed the equilibrium thermal
fluctuations which follow from Gibbs' theory. A simple isolated system—as
represented by an isoenergetic numerical simulation of a typical textbook
mechanics problem—cannot exhibit this type of irreversibility. The fact
that simple isolated systems *can* just as well run backward is the gist of
Loschmidt's reversibility objection to Boltzmann's H Theorem. The additional fact that such systems must eventually *recur*—indicating cyclic
rather than monotonic behavior—was Zermélo's recurrence objection, also
mentioned in Chapter 2. Computer simulations of irreversible behavior,
of the kind seen in real life, require either (i) so many degrees of freedom
that the decay of the initial state persists for a while, allowing an accurate
transient analysis, or (ii) thermal links to the outside world sufficient to lose
the extraneous information generated by nonequilibrium decay processes.

Lost information becomes heat when it is transferred to the outside world. When "information" is used in a technical sense, it refers to the precision—in binary bits—with which a calculation is specified. The same information concept can also be used to quantify the precision of experimental measurements. Without *complete* knowledge of the dynamical state the dynamics cannot be *exactly* reversed. Complete knowledge is only possible in a space with finite information content, such as the integer-valued coordinate space inhabited by Levesque and Verlet's bit-reversible trajectories. For an "open" system, coupled to the external world, and not isolated, there *can* be a time-symmetry breaking making a time reversal *impossible in principle*. This is a consequence of the relative stability of processes which destroy, rather than create, information.

In the real world, apart from equilibrium microscopic fluctuations like Brownian motion, heat invariably flows from hot to cold. This is one way, attributed to Clausius, of stating the Second Law of Thermodynamics. Similarly, macroscopic differences in chemical potential or velocity, unless stabilized by external fields or forces, invariably disappear. This inexorable degradation of gradients is quantified by the entropy increase of the macroscopic Second Law. It has nothing to do with the initial conditions, computer algorithms, or the microscopic details describing the past history of the system. It is a simple phenomenological description of inevitable decay, based on observation.

Despite all this natural irreversibility, the fundamental conservation equations describing real continuum flows are neutral regarding reversibility. The *continuity equation* is the basis of this neutrality. It is the partial differential equation expressing the flow of a conserved quantity in the presence of gradients. Let us apply it to the conservation of mass. The derivation is simplest in a fixed "Eulerian" frame. In this frame the time-rate-of-change of a smoothly-continuous mass density at a fixed location, $\partial \rho / \partial t$, must exactly balance the net flow of mass toward that location:

$$\partial \rho / \partial t = -\nabla \cdot (\rho v).$$

This is the Eulerian form of the continuity equation. An alternative description, in the comoving Lagrangian frame, gives an equivalent form of the differential equation for $\dot{\rho}$, the "Lagrangian" time derivative of density *following* the flow:

$$\dot{\rho} = -\rho \nabla \cdot v \longleftrightarrow \partial \rho / \partial t = -\nabla \cdot (\rho v).$$

The use of Eulerian and Lagrangian continuum simulation techniques is detailed in Chapter 6. Evidently reversing time in the continuity equations simply changes the sign of the time derivatives, matching the sign changes of the velocity on the righthand sides of the two versions of the continuity equation. Either continuity equation is precisely time-reversible.

The Eulerian and Lagrangian "equations of motion" are derived in a similar way. Equate the rate of change of ρv, the momentum density at a fixed location, to the negative divergence of the corresponding flux. The comoving Lagrangian momentum flux defines the pressure tensor P. The Eulerian momentum flux contains the additional convective contribution, $\rho v v$. The resulting equations of motion are:

$$\partial(\rho v)/\partial t = -\nabla \cdot (\rho v v + P) \longleftrightarrow \rho \dot{v} = \rho[(\partial v/\partial t) + v \cdot \nabla v] = -\nabla \cdot P.$$

Here the lefthand sides are *even* functions of time, as can also be the right, *provided* that the pressure tensor is an *even* function of the fluid velocity $v(r, t)$. Evidently the time-reversibility of the equations of motion depends solely on the time-reversal properties of the pressure tensor.

The Eulerian and Lagrangian "energy equations" follow similarly, by expressing the energy change in terms of (i) work done and (ii) heat transfered. In the comoving frame the energy equation has the simple form

$$\rho \dot{e} = -\nabla v : P - \nabla \cdot Q,$$

where Q is the heat-flux vector—the conductive flow of energy in the comoving frame, per unit area and time. The Lagrangian energy equation is the dynamical analog of the First Law of Thermodynamics. In the fixed Eulerian frame, the convective contributions to the energy flux lead to a more complicated, but fully equivalent, Eulerian energy equation:

$$\partial(\rho[e + \tfrac{v^2}{2}])/\partial t = -(\nabla v : P) - \nabla \cdot (\rho v[e + \tfrac{v^2}{2}] + Q).$$

In microscopic molecular dynamics, the atomistic equations of motion are generally time-reversible. The instantaneous microscopic pressure tensor, discussed in Chapter 3, *is* an *even* function of the particle velocities while the heat flux vector is *odd*. The situation in continuum mechanics is typically very different. There, both nonequilibrium fluxes are generally *irreversible*. The dissipative Newtonian pressure tensor is an *odd* function of velocity and Fourier's heat-flux vector is *even*. We discuss the linear form of the macroscopic irreversible laws, together with their theoretical bases in Gibbs' theory, in the next two Sections.

4.2 The Phenomenological Linear Laws

In the absence of dissipative constitutive relations, most equations of state would lead to unstable continuum behavior. The increase of wave velocity with density eventually leads to catastrophic discontinuous shockwaves unless some offsetting smoothing effect, such as viscosity, heat conductivity, or plasticity, is included. The simplest stable continuum model is a "Newtonian" (linearly viscous) fluid, one exhibiting dissipative (irreversible) behavior. Newtonian fluids have shear stresses which are *odd* functions of the stream velocity v. The Newtonian pressure tensor for a two-dimensional or three-dimensional continuum fluid is:

$$P_{\text{Newton}} = I[P_{\text{eq}} - \lambda \nabla \cdot v] - \eta[\nabla v + \nabla v^t] \; ; \; \lambda \equiv \eta_V - (2\eta/D).$$

I is the "unit tensor", with diagonal elements of unity and off-diagonal elements of zero. D is the dimensionality of the system, either 2 or 3. ∇v^t is the "transpose" of ∇v, with

$$\{(\nabla v^t)_{ij} \equiv (\nabla v)_{ji}\},$$

and is included in order to guarantee the symmetry of the pressure tensor, $P_{ij} \equiv P_{ji}$. The equilibrium pressure P_{eq} as well as the bulk and shear viscosity coefficients $\{\eta_V, \eta\}$, are state functions which are independent of the past history of the system and the direction of time. Newtonian viscosity enhances stability by spreading shockwaves over a width of approximately one fluid mean free path. From a macroscopic point of view this width is negligibly small. For this reason a much larger "artificial viscosity", spreading shockwaves over the small-scale numerical resolution length, has to be used to stabilize macroscopic simulations against shockwaves.

In addition to problems involving sound waves and shockwaves continuum simulations can treat mass and energy flows. These are relatively *slow* flows, governed by the stabilizing processes of mass and heat diffusion. The diffusive flow of mass can be described by Fick's Law and the flow of heat by Fourier's Law, which defines the heat conductivity, κ in terms of the heat current responding to a temperature gradient:

$$Q \equiv -\kappa \nabla T.$$

The three phenomenological linear diffusive laws, Fick's, Newton's, and Fourier's, are among the oldest quantitative descriptions of material behavior. All of them provide irreversible solutions to macroscopic initial-value problems. A major accomplishment of Maxwell and Boltzmann's kinetic theory was the demonstration that the *same* linear diffusive laws, found phenomenologically, follow also from an accurate theoretical analysis of low-density gas dynamics. Einstein followed this up with a careful analysis of Brownian Motion, relating two phenomenological irreversible coefficients, fluid shear viscosity, and the Brownian particle's diffusion coefficient.

4.3 Microscopic Basis of Linear Irreversibility

Half a century passed before Green and Kubo made a major advance beyond Maxwell and Boltzmann's gas theory. They related the macroscopic irreversible diffusive transport processes to the decay of the microscopic equilibrium fluctuations described by Gibbs' ensemble theory. Green and Kubo evaluated the effect of perturbations to Gibbs' equilibrium ensembles. The perturbations were specially chosen to *drive* corresponding nonequilibrium mass, momentum, and energy currents. In the linear-response regime, where the effects are proportional to the perturbations, the phenomenological linear diffusive laws of Fick, Newton, and Fourier result. Because Green and Kubo's approach is not at all restricted to gases, but applies equally well to liquids and solids, it provides a *general* understanding of the macroscopic phenomenological laws. These represent, on the one hand, the dissipation found in nonequilibrium steady flows, and on the other, the rate at which equilibrium fluctuations decay. This understanding provides an effective response to Krylov's criticism of Gibbs' theory as a basis for treating irreversibility.

Linear-response theory showed that there is no special new idea that is required in order to treat linear macroscopic irreversibility. It was only necessary to develop computer algorithms with appropriate perturbations, or boundary conditions. The algorithms were developed for use in nonequilibrium molecular dynamics simulations in the 1970s. All three linear laws were reproduced in this way. The transport coefficients could all be expressed as the integrated decays of corresponding equilibrium fluctuations. In the 1970s steady-state thermostats provided an alternative representation of the *same* coefficients, obtained from *energy balance*, in the presence of steady fluxes responding to appropriate driving forces. Despite all this conceptual progress, the early computer simulations, designed to evaluate the microscopic Green-Kubo fluctuation formulæ, were disappointing and confusing, containing as they did independent factor-of-two errors in the heat conductivity and bulk viscosity. Dynamical steady-state simulations were carried out soon afterward, and turned out to agree fairly well with the corrected Green-Kubo results.

About fifteen years ago our understanding of nonequilibrium systems was enriched by a further conceptual advance. New thermomechanical motion equations, incorporating fully deterministic and time-reversible "ergostats", "thermostats", and "barostats", were developed. These new motion equations were capable of controlling a wide variety of mechanically-driven and thermally-driven nonequilibrium systems. The additional thermostat forces, always of the form $\{-\zeta p\}$, came from many disparate sources: Gauss' (Least-Constraint) Principle, Hamilton's (Least-Action) Principle, the Green-Kubo relation for heat flow, and the Nosé and Dettmann Hamiltonians. The main ideas are described in Section 2.8. The older irreversible Langevin approach to temperature control is described in Section 2.9. The new thermostat forces made it possible to simulate transport processes accurately in relatively small systems of up to a few hundred particles.

Now, a generation later, and with our greatly increased computational capability, it is certainly feasible to simulate transport processes in a more "realistic", or at least conventional, way. We *could* simulate a very large isolated Newtonian system incorporating large gradients in its initial conditions. By studying the decay of these gradients we *could* find the corresponding transport coefficients, including both nonlinear effects and number dependence. "Escape-rate theory"—a recently-developed theoretical approach, is a variation of this idea. Escape-rate theory analyzes the decays of equilibrium ensembles, from which mass, momentum, and energy are allowed to *escape* through an open system boundary. *But*, given the relative simplicity of steady-state simulations, there is no convincing reason to characterize transport processes through more cumbersome studies of transient decay.

4.4   Solving the Linear Macroscopic Equations

The "equation of motion" for a continuum is just a macroscopic form of Newton's Second Law of Motion. It gives the local acceleration at a point, $\dot{v}(r, t)$ in terms of the local density ρ and the divergence of the pressure tensor P (which is the negative of the "stress tensor" σ):

$$\rho\dot{v} = -\nabla \cdot P = \nabla \cdot \sigma.$$

Heat flow makes no explicit contribution to the motion equation, though in general the pressure tensor depends on the local temperature. The equation of motion has two linear limits suited to theoretical analysis. Weak long-wavelength disturbances involve negligible heat transfer and viscous dissipation. They are adiabatic and nearly reversible. A systematic expansion of the equation of motion, combined with the continuity equation, $\dot{\rho} = -\rho\nabla \cdot v$, leads to the time-reversible "wave equation":

$$\partial^2 v/\partial t^2 \simeq (-1/\rho)(\partial/\partial t)(\nabla \cdot P) \simeq -(\partial P/\partial\rho)_S(1/\rho)(\partial\nabla\rho/\partial t) \longrightarrow$$

$$(\partial^2 v/\partial t^2)_r = c^2\nabla^2 v \; ; \; c^2 \equiv (\partial P/\partial\rho)_S.$$

A natural description of the solution is in terms of lossless density and pressure waves, "sound waves", which propagate at the *sound* velocity c.

If we ignore the density changes leading to such wave propagation, and assume instead an *incompressible* shear flow with constant viscosity, constant density, and no velocity divergence, the residual motion, with velocities much less than c, is governed by the *shear* viscosity:

$$\rho\dot{v} = \rho[(\partial v/\partial t)_r + v \cdot \nabla v] = \eta\nabla^2 v \; ; \; \nabla \cdot v \equiv 0.$$

The nonlinear term $\rho v \cdot \nabla v$ does not necessarily vanish in the incompressible case. In a plane incompressible shear flow, with $v_x \propto y$, $v \cdot \nabla v$ has an x component proportional to the gradient of the "Reynolds' stress" $\rho v_x v_y$. If the nonlinear terms are ignored, "laminar flow" results, governed by a time-irreversible diffusion equation:

$$(\partial v/\partial t)_r = (\eta/\rho)\nabla^2 v.$$

The lefthand side is *even* and the righthand side is *odd* on time reversal. Similar simplifying assumptions for the "energy equation"—negligible velocity and density gradients with constant heat capacity C and constant heat conductivity κ—lead to another version of the same diffusion equation, but for temperature rather than velocity:

$$(\partial T/\partial t)_r = D_T \nabla^2 T = (m\kappa/\rho C)\nabla^2 T.$$

The thermal diffusivity D_T is dimensionally a diffusion coefficient. This diffusion equation is also intrinsically *time-irreversible*, with the lefthand side *odd*, and the righthand side *even*, in time.

There is no special difficulty in solving diffusion equations for *increasing* time. Either a truncated Fourier analysis or a grid-based numerical approach can be used. But any attempt to work *backward* in time, to find the past history of a system obeying the diffusion equation, is doomed to failure by a short-wavelength instability. To see this, consider the behavior of a single Fourier component of the temperature, satisfying the diffusion equation. If the wavelength is $\lambda = (2\pi/k)$, the corresponding temperature perturbation decays in a time of order λ^2:

$$\delta T(k) \propto \sin(kx)e^{-t/\tau} \longrightarrow \tau = (\lambda/2\pi)^2/D_T = 1/(k^2 D_T).$$

The decay time approaches zero for small wavelengths. The *time-reversed* diffusion equation, which would, if it made sense, determine the past from the present state, exhibits a corresponding exponentially-unstable *divergence* for *any* perturbation:

$$\partial \delta T/\partial t = -D\nabla^2 \delta T \longrightarrow \delta T(k) \propto \sin(kx)e^{+t/\tau}.$$

Thus the diffusion equation cannot be followed backward in time without artificially suppressing the short-wavelength components with their unbounded growth rates.

4.5   Nonequilibrium Entropy Changes

Because entropy furnishes the version of equilibrium thermodynamics closest to Gibbs' statistical mechanics, there is a strong motivation to generalize the concept to nonequilibrium systems, as did Onsager in his own linear-response studies. But there is no straightforward way to make the generalization. *Changes* in the equilibrium entropy, a state function, can only be characterized for processes linking two or more *equilibrium* states together. Bridgman* pointed out that the thermodynamic entropy is undefined for all but the simplest homogeneous fluids or perfect solids. Systems with crystal defects or long-lived chemical complexes cannot generally be created by performing thermodynamically-reversible operations on simple equilibrium states. Bridgman emphasized a subject which he knew well, the irreversible plastic flow of metals. His conclusion was this:

> *The final abiding place of the entropy of irreversibility is*
> *in the heat reservoir surrounding the working body ... .*

A *global* form of the Second Law of Thermodynamics states that the total entropy increases with time. Two prototypical systems for the study of steady nonequilibrium flows are shown in the Figure. The steady homogeneous shear of a Newtonian viscous fluid, with strain rate $\dot{\epsilon} = dv_x/dy$ and shear viscosity η, requires a stress, $\sigma_{xy} = \eta\dot{\epsilon}$ to sustain the flow. The rate at which work is done by this viscous force is

$$\sigma\dot{\epsilon}V = (\eta\dot{\epsilon}^2)V = (\sigma^2/\eta)V.$$

Dividing by the temperature gives the corresponding rate of entropy increase of the system. It is significant that the entropy production is *quadratic* in the deviation from equilibrium. If it is assumed that the flow is steady, with heat flowing to the boundary accounting for the work done, then the surrounding heat reservoirs incorporate all of the entropy increase, just as Bridgman suggested.

In a steady heat flow, with a heat flux Q through a cross-section A between two heat reservoirs separated by a distance L, the total rate of entropy increase *of the reservoirs* is:

$$(QA/T_{\text{cold}}) - (QA/T_{\text{hot}}) \simeq \kappa V |\nabla \ln T|^2,$$

where the total system volume V is AL. Again the entropy production is quadratic in the deviation from equilibrium. If we attempt to apply a similar global reasoning to the *system* entropy, evaluating its change due to heat transfers *from* the hot-boundary heat source and *to* the cold-boundary sink, the result would be a corresponding *decrease* of the system entropy by the same amount, $\dot{S}_{\text{system}} \simeq -\kappa V |\nabla \ln T|^2$. This decrease is traditionally offset by fiat, attributing an additional "entropy production" *to* the

---

*See his article discussing "Generalized Entropy" in the 1950 Reviews of Modern Physics.

conducting fluid, to ensure that its steady state shows no entropy change
whatever. The same general idea can also be applied to shear flow, where
the system entropy would likewise *seem* to decrease, due to the loss of dissipated shear work, in the form of heat, at the upper and lower boundaries.

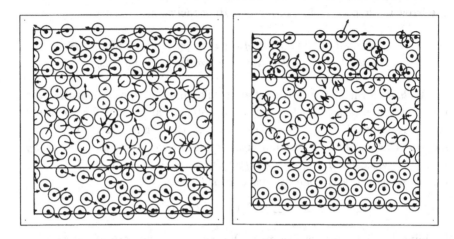

Figure 4.1: Prototypical simulations of two-dimensional shear flow (left)
and heat flow (right) driven by two nonequilibrium reservoirs.

To offset entropy decreases due to both these effects, shear stress and
heat flow, the compensating density of "entropy production" in a fluid near
equilibrium, with Newtonian shear viscosity η and Fourier heat conductivity
κ, is set equal to:

$$\dot{s} = (\dot{S}_{\text{prod}}/V) = (\eta/T)\dot{\epsilon}^2 + \kappa|\nabla \ln T|^2.$$

The entropy production needed to keep the steady-state system entropy
constant is quadratic in the velocity and temperature gradients, for small
deviations from equilibrium. It becomes substantial at strain rates comparable to particle collision frequencies and for energy gradients comparable
to the interparticle forces.

How are these prototypical entropy-producing flows related to the Second Law of Thermodynamics? For steady heat flow the *first* of the four
informal Second Law statements listed in Section 2.7—"Heat cannot flow

from a colder body to a hotter one"—implies that both the conductivity κ and its contribution to the entropy production must be positive. The latter result is in accord with the *second* statement—"entropy must increase"—as well as the *last*: $\oint dQ/T < 0$. For steady shear flow this last statement implies that heat is produced *within* the sheared system, and flows *out*. The positivity of the shear viscosity η and its contribution to the entropy production is also required by the *third* statement: "no *cyclic* process can convert heat entirely to work". The "entropy production", *within* the conducting or shearing fluid, although *consistent* with the Second Law of Thermodynamics, has no operational significance. Neither entropy, nor its internal production rate, can be measured, for nonequilibrium states. Bridgman and Occam would both suggest we shun nonequilibrium entropies.

4.6   Fluctuations and Nonequilibrium States

In *macroscopic* continuum mechanics, a nonequilibrium state is easy to recognize. There are gradients, $\{\nabla v, \nabla T\}$ which cause corresponding nonequilibrium fluxes in the stress tensor and heat flux vector. From the microscopic view these nonequilibrium state variables $\{\nabla v, \nabla T, \sigma, Q\}$ all correspond to averages—space averages, time averages, or ensemble averages. To characterize the "nonequilibrium" character of a particular *microscopic* N-body state is not always so clearcut. It requires significant deviations, over and above the relatively large mechanical and thermal fluctuations $\propto N^{-1/2}$. In the laboratory, where the number of particles is of order 10^{24}, sensible nonequilibrium deviations can be many orders of magnitude smaller than those in computer simulations.

Probabilities for *equilibrium* fluctuations can be calculated on the basis of Gibbs' ensemble theory. At equilibrium the mean values of all the small-system fluctuations, $\{\Delta E, \Delta V, \Delta P, \dots\}$ vanish, just as do all the *first* derivatives of the entropy with respect to these fluctuations. Because the *second* derivatives vary as $1/N$, macroscopic reservoir contributions are negligible compared to those of a microscopic small system. Straightforward thermodynamic manipulations of the second derivatives[†] lead to the small-system probability density for deviations from equilibrium:

$$(f/f_{\text{eq}}) \longrightarrow e^{-(\Delta P)^2/[2kT(\partial P/\partial V)_S]} \, e^{-(\Delta S)^2/[2kC_P/m]}.$$

[†]These are given in Section 114 of Landau and Lifshitz' excellent *Statistical Physics*.

C_P is the constant-pressure heat capacity. Thus, in accord with the Central Limit Theorem, Gibbs' formulation leads to mean-squared pressure and specific entropy fluctuations inversely proportional to system size:

$$\langle (\Delta P)^2 \rangle = \rho c^2 (kT/V) \propto 1/N \propto (kC_P/m)/N^2 = \langle (\Delta S/N)^2 \rangle.$$

Consider a continuum computer simulation of water undergoing steady shear. An $L \times L \times L$ element of volume must be sufficiently large that the pressure fluctuations just estimated,

$$\langle |\delta P| \rangle = \rho c (kT/mN)^{1/2} \propto L^{-3/2},$$

can be ignored relative to the nonequilibrium shear stress, $\eta \dot{\epsilon}$. At the same time, a useful deterministic continuum description requires that a computational volume element not be so large as to exhibit turbulence. Turbulence can occur whenever τ_{shape}, the characteristic time for a volume element to change *shape* $1/\dot{\epsilon}$, is much less than τ_{decay}, the time $L^2/\nu = L^2\rho/\eta$ required for the strain rate $\dot{\epsilon}$ to *decay*. The ratio of these two times is "Reynolds' Number":

$$(\tau_{\text{shape}}/\tau_{\text{decay}}) \equiv R \equiv L^2 \dot{\epsilon}/\nu.$$

The phenomenological inequality $\dot{\epsilon}L^2 << 2000(\eta/\rho)$ describes the range of strain rates for which turbulence can be ignored within a region of side-length L.

The two restrictions just discussed—sufficiently *large* shear stress with sufficiently *small* strain rate—govern the useful range of cell (or "zone" or "element") sizes for realistic computer simulations of viscous water. Evidently L cannot exceed a meter, and the strain rate cannot lie below 0.001 hertz in a useful description. For a typical zone size of one centimeter, the strain rate would have to lie in the range from 0.03 to 20 hertz. Atomistic computer simulations are limited by the same formal considerations, but on a much smaller scale. For a "large" micron-sized atomistic simulation, the strain rates for which shear stresses are large enough to measure, but not so large as to induce turbulence, range upward from 3×10^7 hertz, being limited above by atomic vibration frequencies. Thus the regions in which the microscopic and macroscopic approaches can overlap is quite limited. The two approaches are intrinsically different.

A simulation which collectively exchanges work or heat with its surroundings exhibits a *global* deviation from thermomechanical equilibrium.

A simulation with a local heat flux or shear stress exceeding thermal fluctuations by a wide margin exhibits a *local* deviation from equilibrium. An alternative characterization of nonequilibrium states—using local temperature and velocity gradients rather than local heat fluxes and stresses—would require either two-point finite-difference evaluation or an equivalent combination of interpolation followed by differentiation. Both of these approaches to describing nonequilibrium states are incorporated in the macroscopic simulation techniques detailed in Chapter 6.

4.7   Deviations from the Phenomenological Linear Laws

The linear laws governing the diffusion of mass, momentum, and energy are based on the observations that the quantities transported are proportional to the gradients driving them. Doubling the magnitude of the velocity gradient $\dot\epsilon = \partial v_x / \partial y$ in a shear flow doubles the required stress, σ_{xy}:

$$(\dot\epsilon \to 2\dot\epsilon) \implies (\sigma = \eta\dot\epsilon \to 2\sigma).$$

This constitutive relation describes "Newtonian viscosity", which applies over a few orders of magnitude in the strain rate, at fixed system size. Fourier heat conduction is likewise based on the principle that doubling the temperature gradient doubles the heat flux:

$$(\nabla T \to 2\nabla T) \implies (Q = -\kappa\nabla T \to 2Q).$$

In both these examples doubling the departure from equilibrium, $\dot\epsilon$ or ∇T, *quadruples* the dissipation rate and the entropy production.

The validity of the linear laws has occasionally been questioned on formal grounds, based on the observation that Lyapunov instability leads to very large changes in particle trajectories (growing exponentially with time), so that the trajectories for even an infinitesimal perturbation bear no resemblance to their unperturbed cousins. The flaw in this criticism has been reviewed by Ichiyanagi. The exponential growth of perturbations is actually responsible for a *lack* of dependence on initial conditions, *ensuring* the validity of the ensemble approach and *justifying* the linear laws when the nonequilibrium gradients are small. What happens outside the region described by the linear laws?

Shockwaves are the ultimate nonlinear irreversible process, converting the kinetic energy of a moving fluid or solid into heat within a distance of a few mean free paths. The conversion process is highly nonlinear. To the extent that transport coefficients can be defined within the shockwave, these coefficients can likewise vary strongly on the microscopic length scale of the mean free path. A generation ago, atomistic simulations of shockwave structure, described in the next Chapter, established that the deviations from linear transport theory are typically surprisingly small in shockwaves provided that "realistic" short-ranged forces are used. If the corresponding irreversible shockwave process is simulated macroscopically, from a continuum viewpoint, by solving the continuum conservation laws, assuming Newtonian viscosity and Fourier's heat conduction, nearly the same profile results as in the microscopic simulation. Evidently the *linear* approximation to transport, though it cannot be completely correct, suffices for simple fluids. On the other hand, more recent simulations, carried out by Oyeon Kum and me, with the help of my wife Carol, and using somewhat "artificial" weaker and longer-ranged repulsive forces, produced temperature profiles with pronounced *maxima*, corresponding to a *negative* heat conductivity on the hot side of a steady moderately-strong shockwave! For more details see Section 5.8.

Because nonlinear effects are often small, the search for nonlinearity in simple fluids *can* be relatively frustrating[‡], and is typically of very little interest from the point of view of engineering design and analysis. The negative conductivity mentioned above is a challenging problem. Much more complicated constitutive properties *often* arise when the underlying molecular structure is complex. Paint, for instance, displays an interesting coupling between shear and normal stresses—when *stirred* with a rotating rod, it can *climb* the rod, exhibiting a *vertical* force in response to a *horizontal* shear deformation. Alloys with a complex microstructure are able to "remember" their original shape after large deformations. But nonlinear effects like these are usually quite small in simple fluids.

4.8 Causes of Irreversibility à la Boltzmann and Lyapunov

There are two relatively simple explanations for macroscopic irreversibility, one for isolated systems; the other for systems interacting with their surroundings: (i) in *isolated* systems more likely "states" (macrostates) replace

---

[‡]Siegfried Hess has had considerable success. See Mareschal's 1997 review for references.

less likely ones; (ii) in *driven* systems, interacting with their boundaries, fractal attractors, though of negligible measure at equilibrium, are qualitatively more stable than are their time-reversed images in phase space, the repellors. The first of these explanations is Boltzmann's idea. The reversibility of Newtonian mechanics shows that Boltzmann's idea can only be correct for times much less than Poincaré recurrence times and for systems which are noticeably disturbed from equilibrium. These difficulties often lead Boltzmann's most enthusiastic supporters to stress the importance of "initial conditions", sometimes as far back as the "Big Bang"!

The second explanation of macroscopic irreversibility, based on modern chaos theory, avoids the reversibility and recurrence objections leveled at Boltzmann's ideas. Information loss and Lyapunov instability and computer algorithms are essential to this newer approach. The possibility of information loss, through lack of conservation of phase volume, invariably results in the formation of strange attractor objects with nonequilibrium fluxes. Despite the exploration of many example problems, our understanding of these objects is still imperfect. The relative irreversibility of the fractal attractors can be expressed in terms of the lost information required to recover their initial state. The simplest caricature of this information loss is the "Bernoulli Map", in which binary numbers, $0 < B < 1$, represented as strings of 0s and 1s, are shifted to the left, with the leftmost 0 or 1 discarded, introducing a "bit" of uncertainty each time the shift occurs. For example, the binary number 00010110 can come from the left shift of *either* of two digit strings (000010110 or 100010110).

To me, time reversibility plays an important rôle in the "understanding" of irreversibility. Time-reversible equations of motion can lead to either of two steady-state solution types. On a time-averaged basis it is quite evident—as is discussed at length in Chapter 7—that evolving comoving extensions in phase, volumes in "solution space" (phase space for a system described by $\{q, p\}$ pairs of coordinates and their conjugate momenta), must either stay the same or decrease. $\langle \dot{\otimes} \rangle \leq 0$. No other possibility (such as a continually increasing volume) is consistent with stability and a nonequilibrium steady state. If the motion equations are both *time-reversible* and *mixing* and if dissipation is possible, so that the comoving volume can decrease, then *any* solution of the equations of motion leads to shrinking and collapse, resulting in a strange attractor. The time-reversed repellor, although unarguably a *formal* solution of the *same* motion equations, is generally both unobservable and unstable.

4.9   Example Problems

One of the simplest nonequilibrium stationary states is thermal convection. This is because the confining boundaries are motionless and stationary. A confined fluid in a gravitational field, with sufficient heating on the bottom and cooling at the top, can be driven into a variety of nonequilibrium states: (i) simple conduction, (ii) stationary or time-periodic convection, or (iii) wild swirling currents with thermal plumes characteristic of turbulent flows. The departure of this nonequilibrium system from equilibrium is conventionally characterized by the dimensionless "Rayleigh Number",

$$R \equiv (\partial \ln V/\partial T)_P \Delta T g L^3/(\nu D_T),$$

where g is the gravitational field strength and ΔT is the temperature difference, $T_{\text{bottom}} - T_{\text{top}}$. The kinematic viscosity, $\nu = \eta/\rho$, and the thermal diffusivity D_T, both have the same units [meters^2/second].

For a two-dimensional system of width $2L$ and height L, with vanishing velocity on both horizontal boundaries and with periodic vertical boundaries, two stationary rolls first appear at a Rayleigh number of about 1700. For chilly water, at 22° C, with $(\partial \ln V/\partial T)_P = 10^{-4}$/kelvin, cells with $L = 1$ centimeter, 1 meter, and 100 meters, the bottom-to-top temperature differences at this threshhold for motion are about 1, 10^{-6}, and 10^{-12} kelvins in the three cases. Only in the last completely-unphysical 100-meter case, with negligible heat flow and hardly distinguishable from an equilibrium situation, would the microscopic equilibrium temperature fluctuations be comparable to the temperature differences in the macroscopic continuum flow. Thus the fluctuations which often complicate microscopic descriptions and simulations of transport processes are of little relevance for macroscopic systems.

Even this familiar "Rayleigh-Bénard" convection problem is sufficiently complicated to frustrate analytic work, unless relatively unrealistic approximations are included. Lorenz introduced a simplified model of steady convection for which a thorough analysis can be carried out. More realistic simulations, including the effects of compressibility and an exact treatment of nonlinear convective effects, can easily be computed with six-figure accuracy by finite-difference methods described and applied in Section 6.7.1.

Here we describe Lorenz' caricature of thermal convection. We also discuss molecular dynamics solutions of the equivalent atomistic problem, by adding heat reservoirs and a gravitational field to what would otherwise have been an equilibrium molecular dynamics simulation.

4.9.1 *Rayleigh-Bénard Flow* via *Lorenz' Attractor*

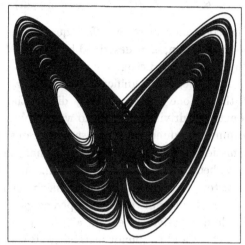

Figure 4.2: Lorenz' Attractor. The calculation shown used 200000 points generated with Runge-Kutta integration. Time step of 0.001. Initially $(x, y, z) = (-7, -11, +13)$. Variation of z with x is shown: $-20 < x < +20$; $0 < z < +50$.

In the early 1960s computer simulation was still in its infancy, restricted to government laboratories and to a few universities with government contracts. In those days it was natural to express the behavior of complicated flows in terms of simple caricature models. Lorenz' model of Rayleigh-Bénard convection is the best known of these. His approximate three-variable description of a system—where the variables $\{x, y, z\}$ represent respectively the instantaneous values of the rotational velocity and the horizontal and vertical temperature gradients—has the form:

$$\{\dot{x} = -\sigma(x - y) \; ; \; \dot{y} = \mathcal{R}x - y - xz \; ; \; \dot{z} = xy - bz\}.$$

Notice that the instantaneous divergence of the (x, y, z)-space "velocity",

$$\nabla \cdot v = (\partial \dot{x}/\partial x) + (\partial \dot{y}/\partial y) + (\partial \dot{z}/\partial z) = \dot{\otimes}/\otimes = -\sigma - 1 - b,$$

is constant and negative. Thus any comoving three-dimensional volume,

$\otimes \equiv dxdydz$, governed by the Lorenz equations, shrinks rapidly toward zero. Lorenz showed that sufficiently large driving (represented by the "reduced Rayleigh Number" $\mathcal{R} \equiv R/R_C$, where R_C is the "critical" Rayleigh number, at which convection begins) leads to the exponential growth of perturbations, "Lyapunov instability". An example trajectory, for the chaotic combination $\{\sigma = 10, \mathcal{R} = 28, b = 8/3\}$, appears in Figure 4.2. For this example the comoving rate of volume change is constant:

$$\dot{\otimes}/\otimes = \langle d\ln(\otimes)/dt \rangle = -\langle d\ln f/dt \rangle = -\sigma - 1 - b = -13.667.$$

The constancy of $\dot{\otimes}/\otimes$ differs qualitatively from the microscopic manybody situation, as is described by Evans, Morriss, and Cohen in their 1993 Physical Review Letter.

It is specially significant that Lorenz' model involves *three* variables. With just one or two ordinary differential equations, and in the absence of external driving, there is no way to obtain chaos. A trajectory confined within a two-dimensional phase space can show no chaos because any deterministic trajectory in a two-dimensional space must either stop or intersect itself, becoming periodic. In three dimensions, where there is no need for a trajectory to intersect itself, chaos can occur, as it does in Lorenz' model.

Lorenz' example retains its pedagogical interest today. It shows that a simple dissipative system, in a three-dimensional state space, can exhibit both chaos and Lyapunov instability. But Lorenz' model is a greatly simplified caricature of real thermal convection. Today complete two-dimensional or three-dimensional simulations of fluid dynamics are routinely carried out using a variety of computational methods. Three such continuum methods are discussed in Sections 6.4-6.6. Here we will next consider the thermal convection problem from the perspective of atomistic molecular dynamics, following simulations pioneered by Mareschal, Kestemont, and Rapaport.

4.9.2 *Rayleigh-Bénard Flow with Atoms*

Numerical solutions of three-dimensional continuum problems can involve dozens of variables at each mesh point. In addition to the five required for density, velocity, and energy, the nine required for the stress tensor and heat flux vector, an additional nine for the velocity gradient, eight for the equilibrium pressure and temperature, and their gradients, give a total of 31 variables. Microscopic molecular dynamics requires just three coordinates, three momenta, and three force components for each point mass. It is therefore considerably simpler to solve the ordinary differential equations of molecular dynamics, using special boundaries to enforce an overall temperature gradient on the system. A particle hitting the lower hot boundary is emitted with a vertical velocity chosen from the Maxwell-Boltzmann distribution:

$$f(v_y) \propto v_y e^{-mv_y^2/2kT},$$

where T is the temperature characteristic of the lower boundary. A similar probability density at the top cold boundary, together with periodic boundary conditions at the sides, completes the description of the system. In order for convective flow to occur, it is necessary that the system size be sufficiently large. Such flows closely resemble the idealization shown in Figure 4.3. Mareschal and Kestemont showed that about 5000 particles are required, in two dimensions, for convection to occur. As the temperature gradient is increased, vertical oscillations of the rolls occur. At higher Rayleigh numbers, chaotic plumes, moving from side to side, can be seen. Rapaport was able to model the vertical oscillations by using 50,000 particles in a microscopic molecular dynamics simulation. By way of contrast, a few hundred "smooth particles" are sufficient to simulate convective flow using the macroscopic continuum equations, as is described in detail in Section 6.7 and Oyeon Kum's 1995 PhD thesis (University of California at Davis/Livermore).

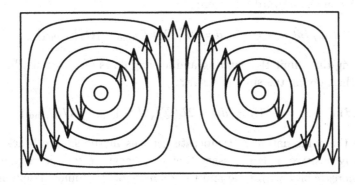

Figure 4.3: Two-Roll fluid circulation in a stationary Rayleigh-Bénard flow. Gravity acts downward. The top and bottom boundaries are cold and hot.

Rapaport carried out a series of atomistic simulations, using hard disks, and found convective flows agreeing well with analogs from numerical continuum simulations. Three different flow regimes were distinguished: (i) conductive flow without convection; (ii) stationary circulation of two counter-rotating vortices; and (iii) periodic vertical oscillations of two rotating vortices. The convective plumes which both experiments and continuum simulations reveal at higher values of the driving temperature gradient are not so easily accessible to atomistic simulations. For a fixed temperature difference and a fixed top-to-bottom gravitational field energy difference, the Rayleigh number varies as $L^2 \propto N$ in two dimensions. The convective plumes, which become prominent at Rayleigh numbers of order 2×10^5 and higher, would require atomistic simulations with hundreds of thousands of particles, a bit too large for modern work stations. On the other hand, continuum simulations on work stations can reach Rayleigh numbers of order 10^8. See Vic Castillo's 1997 papers and his 1999 PhD thesis (University of California at Davis/Livermore).

All of the molecular dynamics simulations just described were carried out with *stochastic* boundaries, choosing new velocities for particles hitting the top and bottom walls from appropriate equilibrium distributions. This makes analysis of the phase-space dynamics relatively complicated. Deterministic friction coefficients would simplify the analysis. As described in Sections 2.8 and 3.4, the instantaneous change in the phase-space probability density f and the extension in phase $\otimes$ depends upon the instantaneous heat flows at the "cold" top and "hot" bottom of the system. We count Q as positive in the upward direction:

$$d\ln f/dt = -d\ln \otimes/dt = 2L[(Q/kT)_{\text{cold}} - (Q/kT)_{\text{hot}}].$$

In the long-time limit the entering and leaving flows of heat must balance one another, giving the simpler result:

$$\langle d\ln f/dt \rangle = -\langle d\ln \otimes/dt \rangle = 2L\langle Q \rangle[(1/kT)_{\text{cold}} - (1/kT)_{\text{hot}}].$$

Because the combination of temperature terms in the square brackets is necessarily positive, it is necessary that Q be positive too, corresponding to the *increase* of f with time and the *decrease* of the comoving extension in phase $\otimes$ characteristic of a strange attractor. This ability to *predict* the direction of the heat flow—from hot to cold—as well as the rate of collapse of the probability density to the strange attractor, is an important advantage of the deterministic thermostats.

4.10 Summary

How is the irreversibility seen in real life to be understood? In most cases the *surroundings* are involved. Without the cat, and the floor, the cup would not shatter. Without the small boy and the rock, or at least the wind, or frogs, fish, or raccoon, there would be no waves dissipating at the water's edge. The *surroundings* act as sources and sinks for both momentum and energy.

Fick's, Newton's, and Fourier's phenomenological macroscopic transport laws describe the simplest nonequilibrium states, the linear mass, momentum, and energy flows which result when sufficiently small gradients are present. Any system, coupled to its surroundings, is subject to perturbing influences which are then amplified by Lyapunov instability. Chaotic expansion and contraction creates and destroys information, making it impossible in principle to retrace the past. The forward evolution invariably seeks out relatively stable attractors while the inaccessible reversed record of this evolution corresponds to an unstable unobservable repellor.

Evidently irreversibility is a consequence of nonlinearity, chaos, mixing, bifurcations, and the overall destruction of information. With nonlinearity, small perturbations can only die, oscillate forever, without change, or diverge. Without chaos and mixing there is no possibility of the memory loss (giving an "insensitivity to initial conditions") which makes a physical description, with observations and simulations, possible. Bifurcations, of the "heads *versus* tails" type, are a simplified model for the time-reversible Lyapunov instability which creates and destroys information and directs systems toward multifractal attractors obeying the Second Law of Thermodynamics.

Chapter 5

Microscopic Computer Simulation

Faith is a fine invention
for gentlemen who see;
But microscopes are prudent
in an emergency.

Emily Dickinson

5.1 Introduction

Microscopic flows begin with individual particle coordinates, velocities, and accelerations. They take place on atomistic space and time scales. To simulate such microscopic flows, using fast computers, involves first of all, a specification of the interparticle forces. These are ordinarily derived from a potential function: $\{F \equiv -\nabla\Phi(\{q\})\}$. The potential energy Φ can be made up of simple pair forces (hard spheres or Lennard-Jones*, for example), and can also include complicated many-body interactions chosen to represent the properties of metals. Over a relatively limited range of thermodynamic states, this force-law approach can lead to useful correlations of data for real materials.

In modeling and understanding real materials it is necessary to estimate the quantum effects which both cause and complicate interparticle forces inferred from experiments. A simpler wholly-classical approach views simulation, for *given* forces, as a data source alternative to laboratory exper-

---

*$\phi = 4\epsilon[(\sigma/r)^{12} - (\sigma/r)^6]$, where σ is the collision diameter and ϵ the interaction strength.

iments and theoretical explanations. It is convenient to imagine that the
chosen force law *defines* a corresponding "material", whether or not a similar material exists in the real world. With the forces chosen, an algorithm
for solving the corresponding motion equations, with or without the influence of boundaries is required. Finally, an analysis of the results is needed.
Because, as kinetic theory indicated, even the simplest interparticle forces,
either impulsive or smooth, are quite sufficient for an understanding of fundamental physics, we will not here touch on the complications which result
from more elaborate models incorporating detailed molecular structure and
long-ranged Coulomb forces. Instead, I emphasize here the simplest choices
of integration algorithms and boundary conditions, as well as methods of
analysis for treating the results gleaned from computer simulations.

5.2   Integrating the Motion Equations

Suppose that we have a representation of the forces in terms of all the
current particle coordinates: $\{F(\{r_0\})\}$. Newton's motion equations can
then be discretized and solved with the "leapfrog algorithm",

$$\{r_+ - 2r_0 + r_- \equiv \Delta t^2 (F_0/m)\},$$

where reasonable accuracy requires that the timestep Δt be considerably
smaller than the shortest vibrational period or collision time. The long-time
error is formally of order $t\Delta t^2$ but can actually increase *exponentially* with
time when Lyapunov instability is present (see Chapter 7). If increased
accuracy is desired, the fourth-order Runge-Kutta algorithm is an excellent
choice. For a set of coupled first-order ordinary differential equations, $\dot{X} = Y(t, X)$ this algorithm takes the form:

$$X_1 = X_0 + (\Delta t/2)Y(0, X_0);$$

$$X_2 = X_0 + (\Delta t/2)Y(\Delta t/2, X_1);$$

$$X_3 = X_0 + \Delta t Y(\Delta t/2, X_2);$$

$$X_{\Delta t} \equiv X_0 + (\Delta t/6)[Y(0, X_0) + 2Y(\Delta t/2, X_1) + 2Y(\Delta t/2, X_2) + Y(\Delta t, X_3)].$$

Because the Runge-Kutta algorithm corresponds to a truncated series expansion of the motion equations' solution, it generally does not reproduce

the time-reversibility of the underlying differential equations. This lack of consistency can be used to estimate the integration error. Because the formal one-step integration error is $\overset{....}{X}\,\Delta t^5/5!$, it is quite feasible to estimate the timestep required to reduce the error to the level of computational roundoff, usually 10^{-14}.

The corresponding timestep Δt is of order $\tau/1000$ where τ is a typical vibration period. Practical calculations can use a timestep ten times larger. It is an advantage of the Runge-Kutta algorithm for first-order ordinary differential equations that it needs no special modifications for nonequilibrium simulations with velocity-dependent forces. It is also possible to generalize the leapfrog approach to deal with these cases. The control variables required to thermostat nonequilibrium steady states are most easily included in a Runge-Kutta integration of the motion equations. In such a case the extent of irreversible heating must first be quantified, and then extracted from the system using the thermostat forces described in Chapter 2.

5.3   Interpretation of Results

In order to draw parallels linking microscopic simulations to macroscopic analyses—thermodynamics and hydrodynamics—mechanical analogs of the pressure tensor, heat flux vector, and temperature are all required. Temperature is basic to nonequilibrium simulations, for any steady-state simulation converts work, or stored energy, to heat. In Chapter 2 we saw that the ideal-gas interpretation of temperature,

$$kT \equiv \langle p_x^2/m \rangle,$$

has both a *dynamical* basis in kinetic theory and a *thermodynamic* basis in Gibbs' statistical mechanics. Unlike the microcanonical equilibrium definition $(\partial E/\partial S)_V$ discussed in Sections 2.8 and 3.5 *this* ideal-gas temperature is useful *far from equilibrium* too. We adopt this ideal-gas definition of temperature under any and all conditions.

The pressure tensor and heat flux vector likewise have two dynamical bases[†]. They can be derived by considering "balance equations" for the momentum and energy within a small volume, or, equivalently, by expressing particle contributions to locally-defined fluxes of momentum and energy,

[†]For derivations of P and Q see Section 5.5 of *Computational Statistical Mechanics*.

using the forces and energies governing the microscopic equations of motion. The volume-averaged pressure tensor P within the volume V is given by:

$$PV = \sum_k [(pp/m)_k + (rF)_k].$$

Each of the tensor components in the first term, a sum over the individual particles $\{k\}$, gives the rate at which the corresponding x or y momentum component is transported in the x or y direction. $P_{xy}V$, the summed-up flow of x momentum in the y direction, includes a contribution $(p_x p_y/m)$ for each particle contributing to the flux. The second term is the dyadic product of the individual particle coordinates and the total forces on the same particle. For *pair* forces, from a pair potential ϕ,

$$\Phi = \sum_{i<j} \phi_{ij},$$

the single-particle (rF) sum can be written as a sum over all distinct particle pairs, with each pair contributing the dyadic $(r_i - r_j)F_{ij}$ to the sum, where F_{ij} is the force on particle i resulting from its interaction with particle j.

The heat flux vector Q is likewise composed of two terms, a convective energy transport and a contribution due to changes in interparticle interactions. The two terms taken together give the complete heat flux. For pair forces the instantaneous, but volume-averaged, heat flux Q within the volume V is given by:

$$QV = \sum_i (pE/m)_i + \sum_{i<j} r_{ij}[F_{ij} \cdot (p_i + p_j)/2],$$

where the individual particle energies $\{E_i\}$ include the comoving kinetic energy as well as half of each interaction energy ϕ_{ij} shared by particle i with other particles j.

In many cases it is desirable to define *local* values of temperature, pressure, heat flux, and the like. A particularly useful description of such variables, with the advantage of two continuous space derivatives, relies on the smooth-particle interpolation method described in Chapter 6. Briefly, local averages of *any* one-particle property X_i, at the location r, are calculated using "weighting functions" $w(r - r_i)$ with two continuous derivatives,

$$\langle X(r) \rangle \equiv \sum_i X_i w(r - r_i) / \sum_i w(r - r_i).$$

The sums include all particles $\{i\}$ within the range of the weighting function. This link between particle and continuum properties is also the basis for a simple numerical method for simulating *continuum* dynamics with *particles*, "Smooth Particle Applied Mechanics", discussed in Chapter 6.

In what follows here we will illustrate the solution and analysis of nonequilibrium motion equations with a simple one-dimensional example, with two-dimensional examples of momentum and energy flows, and with a shockwave simulation. To conclude this Chapter we will work out three nonequilibrium example problems, the Galton Board, and heat flow in both one-body(!) and many-body systems.

5.4   Control of a Falling Particle

An isolated nonrelativistic particle in a gravitational field continues to accelerate for all time, never reaching a steady state. A particle falling in a viscous medium can reach a steady nonequilibrium state in which the frictional force balances the gravitational acceleration. A particle falling in a vacuum, or through a network of fixed scatterers, can also reach a steady state, provided that the motion is controlled, or thermostated, by artificial thermostat forces of the type described in Chapter 2. As an example of artificial control, consider a falling particle from the standpoint of Dettmann's Hamiltonian, the simplest approach to isothermal Nosé-Hoover dynamics, as discussed in Section 2.8. If we set the constants defining the temperature, viscous relaxation time, gravitational field strength, and mass, $\{kT, \tau, g, m\}$, all equal to unity, Dettmann's Hamiltonian is:

$$\mathcal{H}_{\text{Dettmann}} = (p^2/2s) + s[-q + \ln s + (p_s^2/2)] \equiv 0.$$

We will see that the momentum p_s conjugate to s corresponds to the usual time-reversible frictional control variable ζ.

For Dettmann's special choice, $\mathcal{H} \equiv 0$, Hamilton's equations of motion are as follows:

$$\{\dot{q} = (p/s) \; ; \; \dot{p} = s \; ; \; \dot{s} = sp_s \; ; \; \dot{p}_s = (p/s)^2 - 1\}.$$

If we express $\ddot{q}$ in terms of $\dot{q} \equiv v$ and set $p_s \equiv \zeta$, Hamilton's equations take on the usual Nosé-Hoover form:

$$\{\dot{q} = v \; ; \; \ddot{q} = \dot{v} = 1 - \zeta v \; ; \; \dot{\zeta} = v^2 - 1\}.$$

Evidently the stationary dissipative velocity $v = +1$ corresponds to a friction coefficient $\zeta = +1$, so that a particular solution of the equations is:

$$\{q = +t \; ; \; v = +1 \; ; \; p = +e^{+t}/e \; ; \; s = e^{+t}/e \; ; \; p_s = \zeta = +1\} \longrightarrow \mathcal{H}_D \equiv 0.$$

From the physical standpoint, gravitational field energy is converted into heat by the motion, with the corresponding heat extracted by the control force $-\zeta v$ representing a heat reservoir. The divergence of the flow velocity, in (q, v, ζ) space is negative:

$$(\partial\dot{q}/\partial q) + (\partial\dot{v}/\partial v) + (\partial\dot{\zeta}/\partial\zeta) = 0 - 1 + 0 = -1,$$

so that the solution near the *stable* fixed point $(v, \zeta) = (+1, +1)$ is "attractive", with the infinitesimal comoving volume element contracting, $\otimes \simeq e^{-t}$.

This dissipative solution, like the original set of Hamilton's equations for the motion of $\{q, p, s, p_s\}$, is also time-reversible, with the variables $\{\dot{q} = v, p, p_s = \zeta\}$ all changing sign in the reversed motion:

$$\{q = -t \; ; \; v = -1 \; ; \; p = -e^{-t}/e \; ; \; s = e^{-t}/e \; ; \; p_s = \zeta = -1\}.$$

The divergence of the reversed flow's velocity in (q, v, ζ) space is positive:

$$(\partial\dot{q}/\partial q) + (\partial\dot{v}/\partial v) + (\partial\dot{\zeta}/\partial\zeta) = 0 + 1 + 0 = +1,$$

or "repulsive", and corresponds to the instability characterizing the *unstable* "repulsive" fixed point $(v, \zeta) = (-1, -1)$, with $\otimes \simeq e^{+t}$.

This simplest of deterministic time-reversible thermostated problems illustrates the relative stability of a dissipative process converting potential energy, or work, into heat. The evolving flow can be thought of as linking the unstable repellor "source" to the stable attractor "sink". Projected into (v, ζ) space, the ultimate steady-state attractor is the single point $(v, \zeta) = (+1, +1)$. More complicated chaotic problems generate strange attractor-repellor pairs in phase space, with the attractors stable relative to their time-reversed repellor counterparts.

If we had applied a *stochastic* heat reservoir with a constant friction coefficient ζ to this same problem we would have obtained the Langevin equation of motion,

$$\ddot{x} = \dot{v} = 1 - \zeta v + \text{"noise"},$$

with the noise amplitude determining the temperature of the thermostated system. In the simplest case, with the bath temperature much less than that of the falling particle, the noise can simply be ignored and the solution becomes the same as that just obtained from Dettmann's Hamiltonian:

$$\{\dot{x} = v \; ; \; \dot{v} = 1 - v\} \longrightarrow \{x = t \; ; \; v = 1\}.$$

These Langevin equations of motion, and their solution, are *irreversible* (because the friction coefficient is constant), but they can still be "inverted", to calculate the past history of the falling ball. The inverted solution, like that calculated with Dettmann's time-reversible friction, is then unstable:

$$\{\delta\dot{x} = \delta v \; ; \; \delta\dot{v} = \delta v\} \longrightarrow \{\delta x = \Delta e^t \; ; \; \delta v = \Delta e^t\}.$$

Figure 5.1 shows streamlines in (v, ζ) space which correspond to choosing different initial conditions for Dettmann's time-reversible equations. The attractive fixed point at $(+1, +1)$ is always the stable sink, while the unstable fixed point at $(-1, -1)$ acts as a repellor source for the flow.

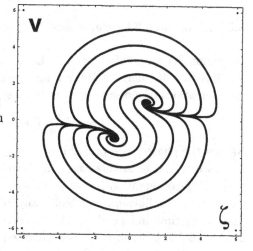

Figure 5.1: Typical trajectories in the (v, ζ) space for the falling particle. The trajectories link an unstable fixed point at $(-1, -1)$ to the stable "attractive" fixed point at $(+1, +1)$.

The *reversible* friction-coefficient approach to simulation has three advantages over the stochastic Langevin alternative: (i) the equations have the useful analytic property of time reversibility, with the friction coefficient(s) $\{\zeta\}$ changing sign in the time-reversed solution; (ii) the results are reproducible, independent of any random number generators or computer idiosyncrasies; (iii) the summed-up friction coefficients are directly related to the rate of phase-space volume collapse—onto a strange attractor—through the compressible version of Liouville's Theorem, detailed in the following Section:

$$-d\ln\otimes/dt = d\ln f/dt \equiv -\sum[(\partial\dot{q}/\partial q) + (\partial\dot{p}/\partial p)] = \sum_p\zeta = \sum_\zeta\#_\zeta\zeta,$$

where $\#_\zeta$ is the number of degrees of freedom thermostated by the friction coefficient ζ. In addition, if the attractor is independent of the initial conditions, which is typical, then the attractor-repellor pairs characteristic of time-reversible motion equations are necessarily ergodic.

Microscopic simulations can only provide suggestive indications of macroscopic behavior, because processor speeds and capacities limit them to tens or hundreds of millions of degrees of freedom. Without this limitation there would be no need for continuum mechanics. Continuum mechanics is likewise limited in the number of equations solved, so that the range of space scales which can be studied is limited to three or four orders of magnitude in two- and three-dimensional Eulerian or Lagrangian continuum simulations.

5.5 Liouville's Theorems and Nonequilibrium Stability

In Section 2.5 we pointed out that Liouville's Incompressible Theorem for the time-rate-of-change of phase volume $\otimes$ is a direct consequence of Hamilton's equations of motion for an isolated system:

$$\mathcal{H} \Longrightarrow \{\dot{f} = 0 \longleftrightarrow \dot{\otimes} = 0\}.$$

Neither the phase-space probability density f nor the comoving phase volume $\otimes$ can change in this case. An extremely useful consequence of this probability-conservation property is an interpretation of isoenergetic thermodynamic equilibrium. Evidently any *equilibrium* phase-space distribution, being time independent, should correspond to a distribution with a fixed probability density and extension in phase. The simplest such distribution is Gibbs' microcanonical distribution, which contains all accessible states of equal energy, all with equal weights.

The intuitive reasoning linking an unchanging distribution to equilibrium could likewise be applied to nonequilibrium *steady states*, which evidently must also be unchanged. Liouville's Theorems can be misleading for that situation because the nonequilibrium distributions are typically fractals, with an overall contracting attractive flow. Liouville's Theorems only apply to differentiable distributions. Applied to *nonequilibrium* situations, for which fractals are appropriate, Liouville's compressible theorem predicts that the fine-grained comoving distribution continually changes. This conclusion would be true for any continuous solution. The rub is that no such solutions exist! The long-time limit is invariably fractal, as is discussed in more detail in Chapter 7.

For the *non*Hamiltonian thermostated equations of motion discussed here, both the density f and the *infinitesimal* comoving volume element $\otimes$ *can* change with time. This follows from the *compressible* form of Liouville's Theorem:

$$\dot{f} = (\partial f/\partial t) + v \cdot \nabla f = -f \nabla \cdot v = -f \dot{\otimes}/\otimes.$$

In the simplest case, using Nosé-Hoover thermostats, with the friction coefficient(s) included as new independent variables, the phase space velocity is $\{\dot{q}, \dot{p}, \dot{\zeta}\}$. The divergence of the flow depends on the friction coefficients:

$$\dot{f} = -f \nabla \cdot v = -f \sum \partial(-\zeta p)/\partial p = f \sum_p \zeta,$$

where the sums are over all thermostated degrees of freedom. As a corollary, the instantaneous "local" identity (where $\otimes$ is infinitesimal),

$$\nabla \cdot v \equiv d \ln \otimes/dt \equiv -d \ln f/dt = -\sum_p \zeta,$$

shows that positive friction causes a loss of phase volume, while negative friction corresponds to increasing volume. From this simple observation it is clear that any stable stationary state must correspond to positive friction, overall, and to the shrinkage of phase volume. This conclusion, the mechanical analog of the Second Law of Thermodynamics, became apparent in 1986, on looking at computer-generated phase-space distributions for simple nonequilibrium systems. The distributions revealed the fractal character which we discuss more fully in Chapter 7. Here we will consider applications of this thermostat idea to both shear flows and heat flows.

The phase-space distribution typically "collapses" ($f \to \infty$; $\otimes \to 0$) onto an ergodic multifractal attractor, like the dissipative attractor resulting from the time-reversible map considered in Section 2.11. Figures 5.7-5.9, shown with the Galton-Board example problem, at the end of this Chapter, likewise illustrate typical multifractal phase-space distributions. The Galton Board, or "Lorentz Gas" was the first such system to be investigated. It consists of a thermostated mass point driven through a regular periodic lattice of hard elastic disks by a constant external field. The *global* time average of the *local* collapse rate, $\langle d \ln \otimes /dt \rangle < 0$, can be expressed in terms of the time-averaged values of the reservoirs' entropy change $\dot{S}$:

$$\sum \langle \dot{S}_{\text{External}}/k \rangle \equiv \langle -d \ln \otimes /dt \rangle \equiv \langle d \ln f/dt \rangle \equiv -\sum \lambda \equiv \sum \langle \zeta \rangle > 0,$$

where the long-time-averaged Lyapunov exponents, $\{\lambda\}$, can be calculated as is described in Section 7.4.

It seems paradoxical that this unidirectional collapse, onto a fractal attractor, is achieved with *time-reversible* motion equations. The details of the unidirectional behavior found with thermostats can depend upon the precise form the thermostats take. This sensitivity to boundary conditions has macroscopic analogs. It corresponds macroscopically to the dependence of drag coefficients on surface friction. This effect is specially noticeable in the frictional force a flowing fluid exerts on the inner walls of pipes with various finishes and surface roughnesses.

An inverted movie of a far-from-equilibrium numerical solution, with the movie projected backward, would certainly show a violation of the Second Law of Thermodynamics. In such an inverted movie heat would "naturally" flow from cold to hot in the presence of temperature gradients; heat would likewise be converted into velocity *differences* in inverted viscous flows. These artificially-*reversed* phenomena, like unburning papers, or divers reëmerging dry from a pool, make no sense because they contradict our everyday experience. It is noteworthy that such artificially-inverted movies rely on *stored* information. They can be continued backward only to the initial state of the motion—the first frame. Any attempt to integrate the equations of motion further backward in time, beyond the initial condition, by replacing $+\Delta t$ with $-\Delta t$, produces a solution indistinguishable, on the average, from the solution forward in time.

Why is the *reversed* trajectory obtained by changing the signs of all the velocities so bizarre, despite the time-reversibility of the equations? From the physical standpoint, the flux directions are counter to our experience. From the mathematical standpoint, a more detailed answer involves the Lyapunov spectrum and Lyapunov instability of any such time-reversed trajectory. Figure 5.2 shows a typical situation in phase space, where the phase space incorporates also any additional friction coefficients required to control the flow. As time proceeds, an infinitesimal comoving hypersphere centered on a trajectory deforms, to a hyperellipsoid, and rotates as it moves. The logarithmic rates of growth $\{\dot{\delta}/\delta\}$ of the offset vectors $\{\delta\}$

(proportional to the corotating principal axes of the infinitesimal hyper-ellipsoid), define the local Lyapunov exponents. Eventually deformation beyond the linear range of the Lyapunov exponents causes the hyperel-lipsoid to bend—as shown in the third view of Figure 5.2, and later to fibrillate. On a global time-averaged basis, the sum of the linear local Lya-punov exponents is necessarily negative (for attractive "stability", as op-posed to repulsive divergence) and gives the time-averaged rate of collapse of a small comoving hyperellipsoid onto the ultimate multifractal strange attractor.

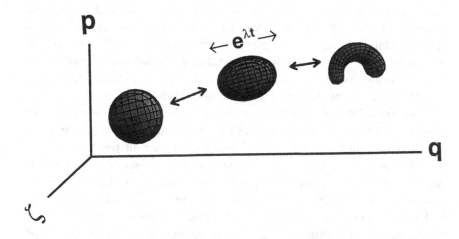

Figure 5.2: Three stages in the schematic time development of a small co-moving phase-space hypersphere. The second view shows the rotation and deformation of the hypersphere to form an evolving hyperellipsoid. The third stage shows the nonlinear bending which, along with shrinking, ulti-mately converts the local hyperellipsoid to a global fractal attractor. The Lyapunov exponents describe the linear, but global, long-time-averaged growth and decay rates parallel to the corotating principal axes of $\otimes$.

Now consider the time-reversed motion. The central trajectory runs *back-ward* and its comoving corotating hyperellipsoid *grows*, on the average, rather than shrinking. The long-time-averaged Lyapunov sum in this re-versed motion is necessarily positive, rather than negative. The sign change corresponds to the sign change from the reversed equations of motion:

$$+\langle d\ln f/dt\rangle_{\text{forward}} \equiv -\langle d\ln\otimes/dt\rangle_{\text{forward}} \equiv -\sum\lambda_{\text{forward}} =$$

$$-\langle d\ln f/dt\rangle_{\text{backward}} \equiv +\langle d\ln\otimes/dt\rangle_{\text{backward}} \equiv +\sum\lambda_{\text{backward}}.$$

A comoving phase volume which *grows* on the average, equivalent to a time-averaged positive Lyapunov spectrum sum, would be *unstable*. The extension in phase would diverge and could not be contained in any bounded region of phase space. In fact, the instability of such a reversed motion would cause that motion to leave the repellor and to seek out the attractor. Any attempt to reverse a dissipative trajectory, short of storing and playing it back precisely, must necessarily fail, with the trajectory soon again seeking out the same attractor which is pursued in the forward direction of time. The severe fractal instability of cold-to-hot heat flow is responsible for time-averaged flows' invariably following the Second Law of Thermodynamics, as is discussed at length in Section 7.8.

This instability necessarily arises because the phase-space strange attractor is far more stable than is its time-reversed image, the strange repellor, upon which f would be a *decreasing* function of time so that the comoving phase volume $\otimes$ would have to increase. The observed and natural asymmetry, with *some* information discarded and lost going *forward* in time, but with additional information absolutely required in order to go *backward* in time, explains the inability of nonequilibrium simulations, or experiments, to recapture the past. Any attempt to do so simply generates another dissipative trajectory, just like those going forward in time. Only by keeping track of the entire trajectory is it possible to regenerate the information-starved reversed trajectory. The lack of symmetry between the forward and backward directions of time, despite formal time reversibility, provides a basis for the analysis of phase-space distributions through unstable periodic orbits.

The introduction of similar constraints, analogous to thermostats, into quantum mechanics can readily be implemented through the use of Lagrange multipliers in the Schrödinger equation. Rather than using an external force to drive the current, a constant current can likewise be maintained by imposing an appropriate constraint on the wave function. Kusnezov discusses an alternative approach, representing quantum thermostats with Gaussian random matrices, in his very readable 1999 paper.

5.6   Second Law of Thermodynamics

For thermostated systems in general there is evidently no analog of the time-symmetric bit-reversible leapfrog algorithm. Whenever the comoving phase volume undergoes local changes, in a space with integer coordinates, a faithful mapping must be locally either one-to-many or many-to-one. Though neither mapping can be exactly reversed, the two possibilities are logically different. A one-to-many mapping requires decisions *creating* information while a mindless many-to-one mapping *destroys* it. This destruction of information, making it impossible to recapture the past, is typical of computer simulations of irreversible processes. Except in very special circumstances—good random-number generators, for instance—any time-periodic mapping, reversible or not, cannot be simultaneously ergodic.

The isokinetic Galton Board problem corresponds to a thermodynamic system which can do work, through a particle's falling in a gravitational field, and generate heat, which is then removed by the isokinetic thermostat forces. This conversion of work to heat is made possible by an essentially infinite energy reservoir in the coordinate-dependent gravitational potential. Consider a many-body nonequilibrium steady state without any macroscopic motion—a system confined between two reservoirs, one hot and one cold. Heat flow, from the hot reservoir to the cold one, leads to an overall increase in the summed-up reservoir entropies:

$$\Delta S_{\text{Reservoirs}} = -(\Delta Q/T_{\text{hot}}) + (\Delta Q/T_{\text{cold}}) > 0.$$

Both the heat-flow example and the thermostated Galton Board are consistent with Clausius' formulation of the Second Law, as described in Section 2.7, with the net change of the reservoir entropies positive.

5.7   Simulating Shear Flow and Heat Flow

We have already considered the macroscopic description of steady nonequilibrium flows, with fixed velocity and temperature gradients. Take a homogeneous "Newtonian" fluid, confined to a volume V (which could be a small volume element), and endowed with a Newtonian shear viscosity η and Fourier's heat conductivity κ. In the presence of a shear strain rate $\dot{\epsilon}$ and a temperature gradient ∇T, the fluid dissipates energy into heat at the rate given by its total "entropy production" $\dot{S}$:

$$\dot{S} = [(\eta/T)\dot{\epsilon}^2 + \kappa|\nabla \ln T|^2]V.$$

Steady flow simulations *defining* the transport coefficients η and κ can be based directly on observations of the shear stress or heat flux induced in a Newtonian simulation region bounded by two thermostated reservoir regions. If the velocity differences driving a shear flow are sufficiently large the generated heat produces an (approximately) parabolic temperature profile, from which the heat conductivity can simultaneously be determined. See Figure 5.3.

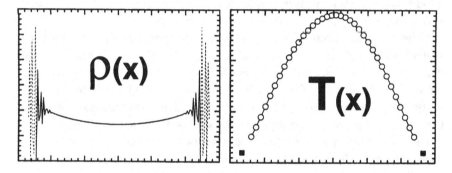

Figure 5.3: Density and Temperature profiles found by Liem, Brown, and Clarke in a stationary simulation of plane Couette flow with thermostated boundaries. See Physical Review A **45**, 3706 (1992) for the details.

The first atomistic transport simulations, based on the definitions of η and κ, and stabilized with thermostating forces, were carried out nearly thirty years ago. They are nicely described in Bill Ashurst's 1974 PhD thesis (University of California at Davis/Livermore). The most straightforward type of implementation used a three-part system, like that of Figure 4.1, page 98, with two reservoirs driving the flow through a central Newtonian region, with *periodic* boundary conditions in the x direction, perpendicular to the systematic nonequilibrium flow of momentum or energy in the y direction. Both the reservoir particles and the Newtonian particles were confined to one of the three subsystems by elastic hard-wall potentials, corresponding to impulsive contact forces acting in the y direction. The equations of motion in the two reservoir regions included "friction coefficients" designed to maintain the flow velocity and temperature within the reservoirs. With the velocities thermostated in both the x and the y directions, as is the usual choice, the reservoir equations of motion are:

$$\{\dot{x} = \dot{x}_0 + (p_x/m) \; ; \; \dot{y} = (p_y/m)\};$$

$$\{\dot{p}_x = F_x - \langle F_x \rangle - \zeta(T)p_x \; ; \; \dot{p}_y = F_y - \zeta(T)p_y\};$$

$$\zeta = \sum(F - \langle F \rangle) \cdot p / \sum p^2/m.$$

The differences $\{F_x - \langle F_x \rangle\}$ ensure that the mean flow velocity $\dot{x}_0$ is unchanged. The friction coefficient ζ is chosen to maintain the desired reservoir temperature, $2kT \equiv \langle (p_x^2 + p_y^2)/m \rangle$. For the simulation of heat flow exactly the same equations can be used, but with the two reservoirs' mean velocities $\{\dot{x}_0\}$ vanishing.

It is important to note that both the Newtonian and the reservoir equations of motion are time-reversible, with the overall strain rate, and the variables $\dot{x}_0, \{p\}, \{\zeta\}$ *all* changing signs along the time-reversed trajectory. Thus any solution of the equations, played backward with these sign changes, would satisfy exactly the same motion equations, to the same numerical accuracy as did the corresponding forward solution. The stability of this hypothetical reversed trajectory is different. The reversed trajectory corresponds to an unstable unobservable repellor flow, as will be considered in detail in Chapters 7 and 8.

Consider the example problem shown in Figure 5.4. It is a computational model of a laboratory viscometer. A many-body fluid, accelerated downward by a gravitational field and slowed by the viscous resistance of a heat-conducting boundary wall, can reach a steady state in which the mechanical acceleration is balanced by viscous drag and the heat transfer to the wall offsets the energy gain from the gravitational field which is dissipated by viscosity. This flow is particularly instructive because it incorporates nonlinear mass, momentum, and energy fluxes which are respectively first-order, second-order, and third-order in the particle velocities. An instantaneous time reversal of the flow, with the gravitational field direction unchanged, reverses the direction of the vertical mass flux ρv_z, the horizontal *conductive* heat flux $Q_r = -\kappa \nabla_r T$, and the vertical *convective* energy flux $\rho v_z[e + \frac{v^2}{2}]$, while leaving the shear stress σ_{rz} unchanged. In the time-reversed situation both the apparent viscosity $\eta \equiv \sigma_{rz}/\dot{\epsilon}_{rz}$ *and the* effective heat conductivity $\kappa \equiv -Q_z/(\nabla T)_z$ would be negative. Heat would flow *from* the walls *into* the fluid, providing the energy for the upward mass flux. This unstable flow would violate the Second Law of Thermodynamics in two ways, by allowing heat to flow from cold to hot while simultaneously converting heat into work.

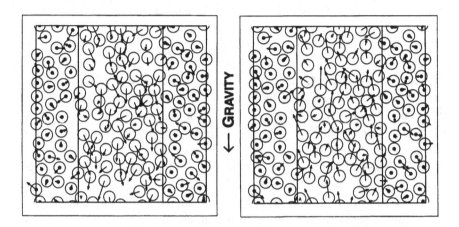

Figure 5.4: Dissipative flow in a model viscometer. Both the *forward* downward flow (left), with the signs of (σ_{rz}, Q_r, J_z) respectively $(+, +, -)$, and the instantaneously *reversed* unstable backward flow (right), with the last two signs reversed, are shown. Gravity is downward, in the negative z direction. Thermostated particles are confined to the two vertical reservoir regions, and are identified here with a central spot.

Relatively small simulations, with perhaps 100 particles in two space dimensions, are sufficient to establish convincing linear velocity or temperature profiles, with transport coefficients determined with errors of order one percent. Spatially-homogeneous modifications of the more complicated three-part systems have also been designed. These simpler systems eliminate the rôle of the boundaries, and reduce the dependence of the results on system size. Such homogeneous simulations have been applied to the irreversible plastic flow of solids as well as to the shear and dilatation of fluids. The fluid viscosities and heat conductivities so found were quite consistent with those from Green and Kubo's linear-response theory based on Gibbs' ensembles. Thus the equilibrium Green-Kubo shear viscosity,

$$\eta = (V/kT) \int_0^\infty \langle P_{xy}(0) P_{xy}(t) \rangle_{\text{eq}} dt,$$

agrees with that obtained by carrying out direct simulations of the corresponding *non*equilibrium systems. The alternative spatially-homogeneous simulations, using periodic boundaries in both directions, have been developed for shear flows and for heat flows stabilized by artificial energy-sensitive fields. This work shows that there are no significant disagreements among these various routes to the simple linear transport coefficients.

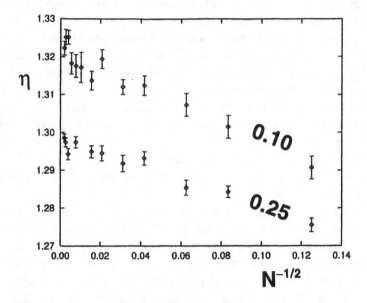

Figure 5.5: Number Dependence of Shear Viscosity in stationary periodic two-dimensional molecular dynamics simulations with a short-ranged repulsive pair potential. The data, all for the same energy per particle, same density, and with two different strain rates, indicate the existence of a finite rate-dependent large-system limiting viscosity, with small-system deviations of order $N^{-1/2}$. The two-dimensional large-system limiting viscosity is thought to have a logarithmic dependence on the strain rate, here $\eta \simeq 1.27 - 0.02 \ln \dot{\epsilon}$. See my 1995 paper with Posch for details.

It is noteworthy that simulations over a wide range of system sizes indicate that the homogeneously-thermostated two-dimensional systems undergoing shear have no special instabilities. At fixed strain rate, density, and energy the shear stress extrapolates smoothly to a well-behaved large-system limit, as is shown in Figure 5.5.

5.8 Shockwaves

A one-dimensional shockwave is arguably the simplest far-from-equilibrium steady state. "One-dimensional" means that the macroscopic averages (stress, energy, heat flux, and the like), as well as the microscopic long-time-averaged distribution functions, depend solely on the spatial coordinate in the direction parallel to the shockwave propagation. One-dimensional shockwaves link two equilibrium states—a "cold" pre-shocked state and a "hot" post-shocked state, with a relatively-localized strongly-nonequilibrium transition region linking them. Because the velocity and temperature gradients can be quite large, with orders of magnitude changes occurring in just a few interatomic spacings, we might expect considerable deviations from the simple Newtonian and Fourier flow laws that apply with small gradients. The possibility of characterizing these deviations, together with the relatively simple geometric requirements, has motivated many shockwave simulations.

The simplest approach to shockwave simulation takes in cold material at one end of the simulation and expels hot material at the opposite end. A proper balance of the two boundary mass currents leads to a stationary position for the shockwave linking the two materials. Entropy *always* increases in the shock process. Typically pressure, density, and temperature increase too, but with the time-averaged mass flux, $\langle \rho v \rangle$, constant. By measuring the velocity and "temperature" and their gradients—where both $(m/k)\langle (v_x - \langle v_x \rangle)^2 \rangle$ and $(m/2k)\langle [(v_x - \langle v_x \rangle)^2 + v_y^2] \rangle$ are possible temperature definitions—the corresponding momentum and energy fluxes, σ_{xx} and Q_x, as well as the effective transport coefficients *within* the shockwave can be computed. These phenomenological nonlinear transport coefficients can then be compared to the *linear* transport coefficients obtained in homogeneous systems. For weak shockwaves the agreement is excellent. For typical pair potentials, even in the very strongest shockwaves, with widths of only a few interatomic spacings, the effective nonlinear transport coefficients lie within thirty percent of their linear counterparts.

In shockwaves the deviations from the linear transport laws are typically not very large. In Figure 5.6 I show an exception to this rule—interesting data for a two-dimensional strong shockwave using *atypical* relatively-weak and long-ranged repulsive forces derived from Lucy's potential,

$$\phi(r < 3) = [5/(9\pi)][1 + r][1 - (r/3)]^3.$$

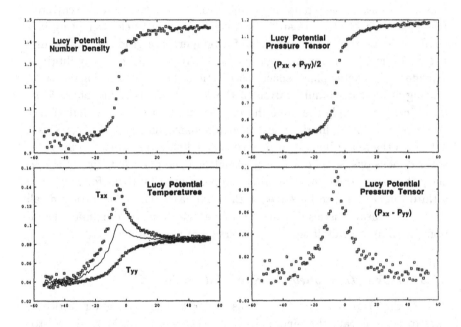

Figure 5.6: Density, temperature, and pressure variations within a dense-fluid shockwave. The Lucy potential was used. Although the heat flux flows from the hot boundary toward the cold boundary throughout, both the transverse and the average temperatures, T_{yy} and $\frac{1}{2}(T_{xx} + T_{yy})$ reach pronounced *maxima* within the shockwave, while the heat flux does not change sign, an aspect which is inconsistent with Fourier's Law.

In this case it is noteworthy that the "temperature" reaches a maximum *within* the shockwave. If Fourier's Law held, even with a nonlinear conductivity, such a temperature profile would imply that the *direction* of the heat flux changes sign within the shockwave. Thus the heat flux and the temperature gradient in this particular simulation are in qualitative disagreement with Fourier's law. This work is described in detail in the 1997 paper by Oyeon Kum, my wife Carol, and me.

5.9 Example Problems

We consider three problems which illustrate the use of deterministic thermostats. First, a single particle falling through an array of fixed scatterers (the isokinetic "Galton Board"); second, a single particle oscillating in a confining potential in the presence of a temperature gradient. These first two problems, involving a single particle driven away from equilibrium, provide multifractal phase-space distributions in few-dimensional spaces, making visualization and analysis of the fractal extensions in phase feasible. Finally, we apply the same thermostat techniques to the simulation of a *many*-body nonequilibrium heat flow in a many-dimensional phase space. Although the many-body heat flow considered here takes place in only two space dimensions, so as to simplify visualizing the particle trajectories, exactly the same equations of motion, supplemented by those for motion in a third dimension, can be solved with equal ease. The three problems all share the characteristics of time-reversible deterministic dynamical equations, with an overall dissipative conversion of work to heat.

5.9.1 *Isokinetic Nonequilibrium Galton Board*

In 1873 Sir Francis Galton used balls falling through a regular lattice of pegs to demonstrate the binomial and Gaussian distributions. An up-to-date review of many aspects of this problem includes the contributions of Carl Dettmann and Harald Posch given in the Bibliography. In the periodic field-free case, worked out in Chapter I, such a ball moves without friction, and eventually comes arbitrarily close to all of its accessible states. The motion is ergodic. An ensemble of such balls, initialized at the center of a large board, eventually spreads out as is described by the *irreversible* diffusion equation. This prototypical illustration of diffusion is simultaneously interesting and manageable, so that the problem has attracted considerable well-deserved attention.

Figure 5.7: 100,000 points $(\alpha, \sin \beta)$ for the isokinetic Galton Board with two moderately-strong fields. Compare the thermostated results at the left to the Newtonian results for the same field strength, shown in Figure 1.5, on page 24. Definitions of the angles $(\alpha, \sin \beta)$ appear at the left. Here the field strengths are $\{mE\sigma/p^2\} = \{0.50, 1.00\}$, where a particle with mass m and speed $|p|$ is scattered by particles of diameter σ. The density of scatterers is $4/5$ the close-packed density. See the 1987 paper by Moran, Hoover, and Bestiale for more details.

A purely mechanical frictionless Galton Board would not function well, for the gravitational energy would eventually cause the falling ball to heat, resulting in a thermal barometer-formula distribution rather than the desired Gaussian distribution at the base. The simplest way to thermostat the falling ball is to constrain it to move at constant speed. The least force required to impose this constraint is a drag force, $-\zeta p$, where ζ is $E \cdot p/2K$, K is the (fixed) kinetic energy, and E is the strength of the accelerating field. The equation of motion is still time-reversible! Both ζ and p change sign in the reversed motion. Similar results, but in higher-dimensional phase spaces, can be obtained with *irreversible* motion equations by using a "coefficient of restitution" (reducing the energy of the moving particle *at* each collision) or by using a linear drag force $-p/\tau$ *between* collisions[‡].

It has been argued that time-reversible equations can show no behavior different in the future from that which they displayed in the past. This argument is reasonable, even true, for isolated systems. But it is certainly *false* for systems driven away from equilibrium, or for systems in the presence of external driving fields. Consider, for simplicity, a ball in a gravitational field. Independent of the initial conditions the ball will eventually fall. In the distant past the ball must have been rising. Thus the future state (falling) and the past state (rising) are quite different. This distinction between future and past presupposes a psychological "arrow of time" making possible the definition of velocity, $v \equiv \dot{r}$.

Figure 5.8 shows also the mirror-image repellor for this same flow, corresponding to the time-reversed attractor trajectory with unstable upward flow, and with the order of the points reversed in time. The inevitable future motion is stable, dynamically, in that the flow collapses onto a strange attractor with a negative Lyapunov sum, $\sum \lambda = -\langle \dot{S}/k \rangle < 0$.

Figure 5.8: 100,000 Attractor (left) and repellor (right) collisions for the isokinetic Galton Board with a relatively-strong field, $(mE\sigma/p^2) = 2.00$. The (unstable and unobservable) repellor states, with $\langle \dot{\otimes} \rangle > 0$, are obtained from the attractor by changing the sign of the velocity and reversing the time-ordering of the points. The abscissa is $0 < \alpha < \pi$ and the ordinate is $-1 < \sin\beta < +1$.

[‡]For references see my paper with Bill Moran—Chaos **2**, 599-602 (1992).

The relative stabilities of the forward and backward motions can differ dramatically if the dynamics is *constrained*, as in the Galton Board problem. Consider an ensemble of moving particles with each particle constrained to move through its own Board at constant kinetic energy. On the average, the future flow is *downward*, and the distribution of collisions is given by the familiar strange-attractor Poincaré sections shown in Figures 5.7 and 5.8.

The conductivity of the Galton Board, $\langle p_x/E \rangle$ varies relatively irregularly with the field strength for low fields. At reduced field strengths above $(mE\sigma/p^2) = 3.69$ interesting limit cycles[§] occur, in which the Lyapunov unstable scattering of the moving particle is overwhelmed by the ordering effect of the field plus thermostat, giving rise to stable quasiperiodic motions.

Because the isokinetic Galton-Board motion equations,

$$\{\dot{x} = (p_x/m) \; ; \; \dot{y} = (p_y/m) \; ; \; \dot{p}_x = F_x + E - \zeta p_x \; ; \; \dot{p}_y = F_y - \zeta p_y\};$$

$$\zeta = Ep_x/2K,$$

are time-reversible, with

$$(+t, +p_x, +p_y, +\zeta) \longleftrightarrow (-t, -p_x, -p_y, -\zeta)$$

in the time-reversed motion, it is easy to generate hypothetical, but wholly unrealistic, states lying in the past. The steps required to generate the past are the same ones detailed in the oscillator example of Section 2.2. A long trajectory segment, obtained by integrating a reversed initial condition "forward" becomes a fictitious unobservable repellor evolution as the result of two operations: (i) reversing the velocities, and (ii) reversing the time-ordering of the points. These steps produced the repellor shown above in Figure 5.8. The unlikely past states at the left—reversed to give the repellor at the right—correspond to trajectories which move upward in the Board, against the field, and with overall *negative* entropy production. This reversed past upward motion is dynamically unstable, corresponding to phase-space expansion and to a positive Lyapunov sum. Time-reversible driven systems *typically* display such a lack in symmetry between future and past, with future states likely, obeying the Second Law,

[§]See the 1987 Journal of Statistical Physics paper by Moran, Hoover, and Bestiale.

and past time-reversed states unlikely, violating it. The Lyapunov spectrum is a key ingredient quantifying the lack of symmetry between the very unlikely, unstable, and unobservable reversed "past" relative to the more-certain relatively-stable future.

In principle, the attractors generated by computer simulation must actually be *periodic* orbits because a finite-precision computation must eventually repeat. For most problems, with roughly 10^{14N} state points for an N-dimensional solution space, the typical orbital period is much too long for an accurate determination. For the two-dimensional Poincaré sections of the Galton Board Figure 5.9 shows the resolution attained with 10^{10}, 10^{14}, and 10^{18} points, corresponding to 5, 7, and 9 digit accuracy in the abscissa and ordinate. The 5-digit simulation is an 11,951-collision periodic orbit or "limit cycle". The singularities separating the various S-shaped regions in the Poincaré sections correspond to boundaries separating scattering collisions with different neighbors of the preceding scatterer.

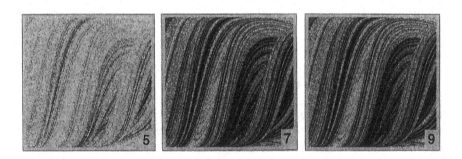

Figure 5.9: Convergence of resolution of typical stationary states for the Galton Board, using five-, seven-, and nine-figure accuracy (left to right) for the abscissa and ordinate values. The field strength is $(mE\sigma/p^2) = 3.00$. Each simulation includes 300,000 successive collisions.

5.9.2 *Heat-Conducting One-Dimensional Oscillator*

The essential singularities associated with the Galton Board problem, and the multifractal phase-space distribution which it generates, complicate the numerical analysis somewhat, and might also complicate theoretical approaches. For that reason I describe a system which is simpler, from the analytic standpoint, a one-dimensional oscillator in a temperature gradient. The motion takes place in a four-dimensional phase space, $\{q, p, \zeta, \xi\}$. The temperature (kinetic energy) of the oscillator is governed by a generalized Nosé-Hoover control of the second velocity moment, $\langle p^2 \rangle$, using *two* time-reversible friction coefficients, ζ and ξ. The extra coefficient is required to ensure ergodicity in the limiting equilibrium isothermal case.

For simplicity I set the oscillator mass and force constant, as well as both thermostat relaxation times, all equal to unity. The thermostat variables, ζ and ξ, control the oscillator's kinetic temperature $\langle p^2 \rangle = kT$ as follows:

$$\{\dot{q} = p \; ; \; \dot{p} = -q - \zeta p \; ; \; \dot{\zeta} = p^2 - kT - \xi\zeta \; ; \; \dot{\xi} = \zeta^2 - kT\}.$$

In the equilibrium case (where the temperature T is constant) these motion equations provide ergodicity with a Gaussian distribution function:

$$kT(q) \equiv 1 \longrightarrow f = (2\pi)^{-2} e^{-(q^2+p^2+\zeta^2+\xi^2)/2}.$$

Ergodicity is most simply checked by propagating a grid of points on the unit hypersphere $q^2 + p^2 + \zeta^2 + \xi^2 = 1$ forward in time, and choosing those points with the most disparate time averages for long-time study. The stationary Gaussian distribution follows directly from the Liouville equation for $(\partial f/\partial t)$, using the Gaussian distribution given above:

$$(\partial f/\partial t) = -\nabla \cdot (fv) \equiv -f\nabla \cdot v - v \cdot \nabla f = -f[-\zeta - \xi] - v \cdot \nabla f =$$

$$-f[-\zeta - \xi] + p(qf) + (-q - \zeta p)(pf) + (p^2 - 1 - \xi\zeta)(\zeta f) + (\zeta^2 - 1)(\xi f) \equiv 0.$$

A *nonequilibrium dissipative time-reversible* variation of this isothermal conducting oscillator model results if the temperature varies smoothly between the limits $1 \pm \epsilon$:

$$kT(q) \equiv 1 + \epsilon \tanh q.$$

As would be expected, so long as the perturbation ϵ is small, the long-time-averaged distribution function is independent of the initial conditions (again, as judged by long-time-averaged values of moments), making it possible to assess the nonequilibrium dissipation and entropy production. The problem is a nice illustration of the difficulties involved in analyzing nonlinear dissipation. The "entropy production" is simply the mean value of the heat extracted from the oscillator, divided by the corresponding temperature:

$$\langle \dot{S}_{\text{external}}/k \rangle = \langle \zeta p^2/mkT(q) \rangle \simeq \langle \zeta \rangle + \langle \xi \rangle.$$

Because the temperature fluctuates, the entropy production is *not* simply related to the rate at which a four-dimensional comoving hypervolume $\otimes \equiv dq\,dp\,d\zeta\,d\xi$ collapses to a strange attractor:

$$\langle d\ln \otimes/dt \rangle = \langle \nabla \cdot v \rangle \equiv \langle -\zeta - \xi \rangle.$$

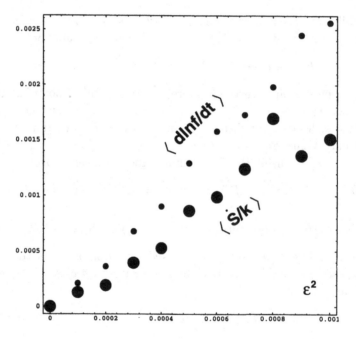

Figure 5.10: $\langle d\ln f/dt \rangle = -\langle \nabla \cdot v \rangle$ (above) and $\langle (\dot{S}/k) \rangle$ (below) as functions of the *square* of the maximum temperature gradient ϵ for a one-dimensional oscillator. The data indicate quadratic relationships, as would be expected from linear-response theory.

For larger values of the maximum temperature gradient ϵ the motion collapses to a limit cycle, a one-dimensional periodic trajectory in the four-dimensional phase space. Numerical solutions for this problem, for a series of equally-spaced values $\{\epsilon^2\}$, show increases in both $(\dot{S}/k)$ and $d\ln f/dt$ which are *close* to quadratic, but not equal, for small ϵ^2. A limit cycle, corresponding to a strictly-periodic oscillator trajectory, becomes stable at maximum gradient values larger than those shown in the Figure. The largest of the four Lyapunov exponents is approximately 0.066 for this problem. See Section 7.10.3, page 193, for details of the far-from-equilibrium case with $\epsilon = 0.25$.

5.9.3 *Many-Body Heat Flow*

To capture the flavor of a serious many-body simulation, while maintaining the graphic advantages of two space dimensions, consider a doubly-periodic four-chamber system (actually an area in two dimensions) with a hot chamber and a cold chamber separated by two Newtonian chambers. Each of the four chambers contains 324 identical particles at an overall number density of unity. The hot-to-cold temperature ratio is three. Pairs of particles interact with a very-smooth short-ranged Lucy potential,

$$\phi(r < 3) = [5/(9\pi)][1 + r][1 - (r/3)]^3.$$

The particle motion in the two thermostated regions is moderated by friction coefficients depending upon the specified equilibrium temperatures of these regions. If we express these equilbrium temperatures by specifying average reservoir kinetic energies, $\{K_0(T)\}$, using Nosé-Hoover control, the corresponding reservoir equations of motion are:

$$\{\dot{x} = (p_x/m) \; ; \; \dot{y} = (p_y/m) \; ; \; \dot{p}_x = F_x - \zeta(T)p_x \; ; \; \dot{p}_y = F_y - \zeta(T)p_y\};$$

$$\{\zeta(T) = \int_0^t [\tfrac{K(t)}{K_0(T)} - 1]dt'/\tau^2\}.$$

The Newtonian motion equations have the same form, but without any frictional terms. Figure 5.11 shows a typical snapshot of the dynamics and a time-averaged temperature profile. Analysis of this problem has established that the hot-to-cold heat flow causes the collapse of the corresponding probability density onto a strange attractor.

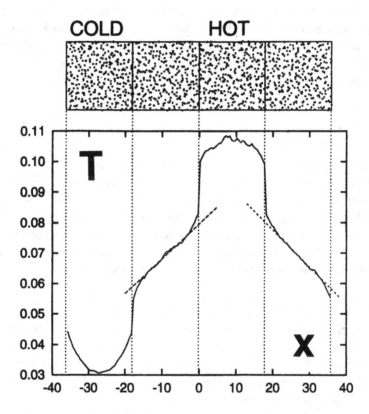

Figure 5.11: Snapshot of the four-chamber 1296-particle simulation of heat flow between two reservoirs and time-averaged temperature profile. For details see "Numerical Heat Conductivity in Smooth Particle Applied Mechanics", by Harald Posch and me, Physical Review E **54**, 5142-5145 (1996).

5.10 Summary

The deterministic time reversibility of Newtonian, Lagrangian, and Hamiltonian mechanics for isolated systems can be carried over to the simulation of thermostated *nonequilibrium* atomistic systems. The nonequilibrium flows which result exhibit a subtle time asymmetry, which was first seen in simulations of the Galton Board problem. Despite the deterministic and time-reversible nature of the equations of motion, the resulting trajectories generate a probability density which is singular everywhere (ergodic *and* multifractal), a *strange attractor*. The long-time-averaged time-rate-of-change of the comoving phase-space volume $\otimes$ is *invariably* negative, corresponding to contraction and collapse, while nearby trajectories *invariably* diverge from one another exponentially, with time. The underlying strange attractor is typically *ergodic*, filling all of the accessible phase space, and corresponds to a dissipative state, as described by the Second Law of Thermodynamics. The time reversibility of the equations of motion guarantees the existence of a topologically similar repellor structure, on which the Second Law is violated. The repellor is less stable than the attractor, and unobservable, in a sense which can be made precise through an analysis of the Lyapunov spectrum, as is explained in Chapter 7. These generic properties of the Galton Board problem were pointed out in my 1986 work with Bill Moran. There is a closely-related paper, describing thermostated motion in a one-dimensional *sinusoidal* potential, which Harald Posch, Brad Holian, Mike Gillan, Michel Mareschal, Carlo Massobrio, and I published at about the same time. The generic features which we found in the computer simulation of the Galton Board and sinuisoidal work, were validated theoretically seven years later, by Chernov, Sinai, Eyink, and Lebowitz.

The time-reversibility of the underlying motion equations is more apparent than real. In the forward direction of time more information is destroyed than is created, with a fractal phase-space distribution formed on a zero-measure phase-space set. Such a rare nonequilibrium flow can only be reversed by storing the entire trajectory and playing it backward.

Chapter 6

Macroscopic Computer Simulation

An art can only be learned in the workshop
of those who are winning their bread by it.

Samuel Butler

6.1 Introduction

For the most part, the microscopic details of macroscopic flows are an irrelevant distraction. By ignoring these atomistic details and focusing directly on macroscopic variables, one might expect a simpler implementation, free of distracting fluctuations. Reduced fluctuations should also lead to improved stability. But the opposite is true. Though the continuum approach *is* a simplification, the partial differential equations of continuum mechanics, with field variables depending upon space as well as time, incorporate a *continuum* of degrees of freedom. Microscopic motion equations are typically well-posed, with well-behaved solutions for as long a time as desired. Relatively simple algorithms are adequate. Macroscopic simulations can exhibit instabilities on a variety of length scales with a corresponding variety of frequencies and growth rates. The wide ranges of the length and time scales, from the free-path atomic-vibration scales of strong shockwaves up to the macroscopic scale, can cause resolution difficulties in addition to the instabilities. The demands of creating and tracking the progress of material surfaces and interfaces add to the complexity of algorithm development in the macroscopic case.

To simulate even the simplest of macroscopic flows we must solve the partial differential conservation equations for the space and time dependence of the density, velocity, and energy:

$$\dot{\rho} = -\rho \nabla \cdot v \; ; \; \rho \dot{v} = \nabla \cdot \sigma = -\nabla \cdot P;$$

$$\rho \dot{e} = \sigma : \nabla v - \nabla \cdot Q = -P : \nabla v - \nabla \cdot Q.$$

There are three prerequisites for solving these flow equations. First, we need to express the stress tensor σ (the pressure tensor is its negative, $P = -\sigma$) and the heat flux vector Q in terms of the time histories of the density, velocity, and energy fields. Then, we must supply initial conditions, along with boundary conditions. It is typical to insist also that the macroscopic constitutive equations be "dissipative", obeying the Second Law of Thermodynamics, ensuring the decay of velocity and temperature gradients, so that *boundary conditions* driving the system away from equilibrium are usually necessary too. Macroscopic dissipation corresponds to the *irreversible* loss of information. That loss makes it impossible to reconstruct the past temperature and velocity gradients which have dissipated differences into internal energy, or "heat". Explicit dissipation—entropy production—ensures that the macroscopic equations are instantaneously irreversible. Entropy inexorably increases, without the need for any time averaging.

We must begin the solution process by reducing the continuum of degrees of freedom to a manageable number. There are several ways to convert the continuum equations into a discrete set: (i) expansion in orthogonal polynomials, such as Fourier's series; (ii) finite-difference approximations; (iii) finite-element approximations; and (iv) particle methods. Of these, particle methods are the simplest to implement. Particle methods closely resemble molecular dynamics, but with the interparticle "forces" directly incorporating the macroscopic constitutive relations for P and Q.

There are in addition several useful approaches to *simplified* approximate forms of continuum mechanics. Relatively *slow* hydrodynamic flows can be treated as *incompressible*. The plastic (shear) flow of metals is also, to a good approximation, incompressible. Large-scale simulations of fluid motion—as in atmospheric turbulence studies—can often afford to ignore millimeter-scale dissipation caused by viscosity and heat conductivity, or can replace the dissipation with simple *ad hoc* models. Except for this last simplification, which makes numerical turbulence studies possible, such restrictions are specially useful in analytic studies, and are not specially relevant to computer simulations.

6.2 Continuity and Coordinate Systems

"Continuum" mechanics incorporates the underlying assumption that the "field variables" describing the continuum vary smoothly in space and time, so that the required space and time derivatives are well defined. With this assumption, the basic partial differential equations can be simply derived. The derivations proceed by considering the flows of the three conserved quantities: mass, momentum, and energy. These can be most simply expressed in either of two coordinate systems, the fixed "Eulerian" frame or the comoving "Lagrangian" frame. The steps are identical to those involved in demonstrating the compressible form of Liouville's Theorem. As an example, consider the "continuity equation" of Section 4.1, which describes density changes resulting from gradients in the mass flux—a vector (ρv) equal to the flow of mass at the location r, per unit area and time:

$$(\partial \rho / \partial t)_r = -\nabla \cdot (\rho v) \longleftrightarrow \dot{\rho} = -\rho \nabla \cdot v.$$

The continuity equation, and its analogs for momentum and energy conservation, are completely independent of microscopic physics and should best be viewed as alternative *macroscopic* descriptions of experience. The underlying ideas are just two: (i) *differentiability* of material properties, with respect to space and time; (ii) *conservation* of mass, momentum, and energy. To illustrate the source of the basic equations let us consider the flow of a material (solid, liquid, or gas) with a mass density $\rho(r, t)$ and a flow velocity $v(r, t)$. The Eulerian derivation is simplest. We choose our coordinates in the laboratory rest frame. Focus on a single square Eulerian sampling zone (often called a "bin", "cell", or "element") with area $\Delta x \Delta y$, centered on the location r and observed for a long enough time interval Δt, that any microscopic fluctuations can be ignored. This zone will experience four mass-flux contributions to its total mass $M = \rho \Delta x \Delta y$:

$$M(t + \Delta t) - M(t) = \Delta t (\partial (\rho \Delta x \Delta y) / \partial t)_r =$$

$$\Delta x [(\rho v_y)_{y - \Delta y/2} - (\rho v_y)_{y + \Delta y/2}] \Delta t + \Delta y [(\rho v_x)_{x - \Delta x/2} - (\rho v_x)_{x + \Delta x/2}] \Delta t.$$

Because (i) Δx and Δy are constant, and (ii) ρv is assumed to be smooth, division by $\Delta t \Delta x \Delta y$, followed by a two-term series expansion of the difference terms gives directly the "Eulerian form" of the partial differential continuity equation:

$$(\partial\rho/\partial t)_r = -\nabla \cdot (\rho v) = -v \cdot \nabla\rho - \rho\nabla \cdot v,$$

describing the change of density at a fixed location in space.

It is equally interesting to ask for the change in density (or other field variables) *following* the motion, just as the *comoving* time derivative in phase space can be evaluated in deriving Liouville's Theorems. The resulting "Lagrangian" time derivative gives the comoving time-dependence along a streamline, following the motion. For density, the Lagrangian form of the continuity equation is:

$$\dot{\rho} \equiv (\partial\rho/\partial t)_r + v \cdot \nabla\rho = -\rho\nabla \cdot v,$$

which is algebraically equivalent to the Eulerian form. In the older texts Lagrangian equations are often used to indicate and describe the motion of a material point as a function of its "original" coordinates at an arbitrary time origin. With the widespread knowledge that many flows involve chaos (and are therefore *exponentially* sensitive to small perturbations) this association of "new" with "old" coordinates assumes more than is reasonable. At present, Lagrangian computer programs, designed to follow particular elements of material, are specially useful when large-scale deformations of several different materials prevail. The equations solved by such a program are the Lagrangian expressions for $\{\dot{\rho}, \dot{v}, \dot{e}\}$. Eulerian computer programs, which are somewhat simpler to write unless it is necessary to keep track of moving interfaces, evaluate the partial time derivatives of these same variables in one or more fixed grids. This approach is natural for a problem with long-time circulation within fixed boundaries, such as Rayleigh-Bénard convection. A Lagrangian treatment of large-scale shears eventually leads to insuperable numerical problems. "Hybrid" methods advance the flow field using the Lagrangian equations and then interpolate the advanced field variables onto a fixed Eulerian grid.

In particular cases (incompressible flow is the best example) "spectral methods" representing the field variables by truncated Fourier series are cost effective. Still another approach to solving the continuum equations is a "particle" simulation method, in which individual representative particles have coordinates, velocities, and energies, which can be different to the field values of these variables at the same location. In the example problems at the Chapter's end, I apply both the finite-difference and particle methods to solving the Rayleigh-Bénard convective heat-flow problem.

6.3 Macroscopic Flow Variables

A solution of the differential flow equations describing a continuum requires a complete description of the "state" of that continuum, including the distribution of mass, momentum, and energy. Predicting future flows of mass, momentum, and energy requires special constitutive laws for the comoving momentum and energy fluxes P and Q. The mass flux ρv is the product of the local density and the fluid's velocity vector. The magnitude of the mass flux corresponds to the flow per unit area across an infinitesimal area normal to the fluid's flow velocity.

While the flow velocity $v(r, t)$ is most naturally measured relative to a fixed laboratory coordinate system, the momentum and energy fluxes (which are respectively tensors and vectors) are conventionally separated into two parts. The "comoving" momentum and energy fluxes—the fluxes relative to a Lagrangian coordinate system moving with the fluid velocity v—are the pressure tensor P and the heat flux vector Q, respectively. The additional "convective" momentum and energy fluxes comprise the "streaming" contributions, the tensor field ρvv for the momentum flux, and the vector field $\rho v[e + \frac{v^2}{2}]$ for the energy flux. In addition to these convective fluxes, energy changes due to the performance of work and the conduction of heat must be included too. These contributions to the energy flux are two vectors, $P \cdot v$ and the heat flux Q, respectively. Their divergences contribute to the rate-of-change of energy density.

Solutions of the continuum equation of motion,

$$\rho \dot{v} = \nabla \cdot \sigma = -\nabla \cdot P,$$

and the continuum energy equation,

$$\rho \dot{e} = \sigma : \nabla v - \nabla \cdot Q = -P : \nabla v - \nabla \cdot Q,$$

require constitutive equations for P and Q. P has both an equilibrium and a nonequilibrium part. The simplest equilibrium equation of state assumes a power-law dependence of pressure on density, $P \propto \rho^{\gamma}$. The most complex equations of state take the form of tables giving the dependence of the pressure on energy and density. The transport properties can be the simple Navier-Stokes and Fourier recipes, including Newtonian shear and bulk viscosities as well as Fourier's linear heat conductivity, or the much more elaborate constitutive relations describing anisotropic solids with tensor heat conductivities and stresses depending upon the orientation and past history of the material. The complexity of the continuum constitutive relations is limited only by the imagination, causality, and the Second Law of Thermodynamics. Relations violating the Second Law result in prompt numerical instabilities because gradients tend to grow, exponentially fast, rather than to decay. I describe two traditional numerical methods for solving the continuum equations in the next two Sections. The smooth-particle method which follows them, in Section 6.6, is an excellent bridge between the atomistic and continuum approaches.

6.4   Finite-Difference Methods

Although it seems most logical to formulate discretized solutions of the *partial* differential equations of continuum mechanics as sets of *ordinary* differential equations for the motion of nodes or the evolution of series-expansion coefficients, it is more usual to treat both time and space as discrete variables. The reason for this is pragmatic. The resulting simulations require less computer time and storage space. In the "doubly-discretized" case it is necessary to correlate the timestep Δt with the spatial increment Δx. In the limit that both Δt and Δx vanish, a proper correlation guarantees stability while maintaining consistency with the underlying partial differential equations. Typically Δt must be less than $\Delta x/c$, where c is the sound velocity, and also less than $\Delta x^2/D$, where D is a diffusion coefficient. Otherwise the solution technique is unstable or inaccurate. In applied continuum simulations it is usual to choose the largest possible timestep consistent with the two kinds of restrictions, because the goal is to reach a particular total time, known in advance.

Straightforward computational techniques can be based on Taylor's series expansions of the continuum equations. A network of sampling points, with a discrete spacing Δx, is carried forward through a series of discrete timesteps, with either a fixed or variable Δt. Some interesting more-elaborate variants can be developed, with the timestep itself varying in both space *and* time. Gradients throughout the network of points are approximated with low-order finite-difference formulæ. Consider the evaluation of the temperature gradient required to drive a heat flow, according to Fourier's Law. The required derivative, dT/dx, for instance, can be approximated by the centered "finite difference":

$$dT/dx \equiv [T_{+\Delta x/2} - T_{-\Delta x/2}]/\Delta x$$

$$\text{error} \propto \Delta x^2 T'''.$$

Improved accuracy can be achieved in either of two ways: by (i) reducing the mesh size or by (ii) including higher-order terms in the finite-difference expression:

$$dT/dx \equiv (9/8)[T_{+\Delta x/2} - T_{-\Delta x/2}]/\Delta x - (1/24)[T_{+3\Delta x/2} - T_{-3\Delta x/2}]/\Delta x$$

$$\text{error} \propto \Delta x^4 T'''''.$$

The extra effort required by higher-order methods is generally better spent increasing the resolution of the simpler second-order approximation. The higher-order expression can be used, from time to time, to estimate errors.

In Sections 4.1 and 6.2 I emphasized that the continuum equations can be written in terms of "Eulerian" derivatives, at fixed points in space, or in terms of the comoving "Lagrangian" derivatives, following the motion. There is a direct computational analog of this distinction. The sampling points, at which field variables like $\{\rho, v, e, \dots\}$ are measured, can be fixed in space or can move with the material. The corresponding forms of the continuum equations then become evolution equations for the field variables at these points. Both choices have advantages and disadvantages. The Eulerian approach makes it difficult to keep track of material interfaces and surfaces but avoids mesh-tangling. The Lagrangian approach takes care of material interfaces in a natural way—they become element boundaries—but can be defeated by large-scale deformations. Turbulent flows or large-scale plastic deformation eventually lead to the failure of Lagrangian approaches. Instead of tracking the motion accurately, the nodes tangle, and require "rezoning"—the construction of a new approximating grid—from time to time.

Occasionally, as in the study of steady shockwave propagation, a velocity different to either the laboratory frame or the material frame is advantageous. The continuum analog of the atomistic shockwave simulation of Section 5.8 can be based on a numerical solution of the continuum equations with the restrictions that the mass, momentum, and energy fluxes are all constant. For a shockwave propagating in the x direction, the corresponding flux components are:

$$\{\rho v_x, P_{xx} + \rho v_x^2, \rho v_x[e + (P_{xx}/\rho) + (v_x^2/2)] + Q_x\},$$

where v_x is the velocity component in the direction of shockwave motion. This one-dimensional problem is relatively elementary and reduces to the solution of two coupled ordinary differential equations [for $\rho(x)$ and $T(x)$] when the continuity equation is used to eliminate the velocity v from the momentum- and energy-conservation laws. In the more usual two- and three-dimensional applications, the finite-difference and finite-element methods are simplest to employ if relatively low-order approximations to the space and time derivatives are used. Constructing and revising the meshes make up a large part of the programming work. A solution of the Rayleigh-Bénard problem using the finite-difference method is the first example problem at the end of this Chapter.

6.5 Finite-Element Methods

Another approach can be based on the notion of a solution with small-scale "local" errors in both space and time, which is only correct in a spatially-averaged sense*. Rather than satisfying the continuity equation exactly, a "weak" solution would ensure that the two (approximate) averages agree:

$$\langle d\ln\rho/dt\rangle_{\text{element}} = \langle -\nabla\cdot v\rangle_{\text{element}},$$

or

$$\langle\dot\rho\rangle_{\text{element}} = \langle -\rho\nabla\cdot v\rangle_{\text{element}},$$

where the angular brackets indicate integration over the element volume and an average over a short time interval Δt. In some cases—elasticity theory and quantum mechanics are the best-known examples—such weak

---

*Such an averaged solution is often called a "weak" solution by mathematicians.

solutions satisfy a variational principle. The true energy of an elastic deformation must lie below the corresponding energy of any approximate deformation with the same boundary displacements. Likewise, the energy of the true ground-state quantum wave function must lie below the corresponding energy of *any* approximate wave function satisfying the same boundary conditions. More generally, as in either version of the continuity-equation example above, finite-element solutions are obtained by insisting that the differential equations' element averages be exactly correct. Again the equations chosen for solution can be either Eulerian, with the element fixed in space, or Lagrangian, corresponding to a comoving element following the flow.

In the Eulerian case the field variables are given throughout each finite element by appropriate interpolation or "shape" functions. In one dimension, for instance, density, velocity, and energy could be linearly interpolated between the two nodal points bounding an element. They could alternatively be represented by "cubic splines", guaranteeing the continuity of the first two derivatives at the element boundaries. In one space dimension the construction of a cubic spline is straightforward. Three adjacent nodal values give the second derivative at each node. Linear interpolation between these, followed by two space integrations matched to the two nearest nodal values, provides the cubic interpolating function. Use of this idea in solving the Rayleigh-Bénard problem requires very little additional computer time and reduces errors in the flow field's kinetic energy by about a factor of three. See Vic Castillo's 1999 PhD thesis (University of California at Davis/Livermore) for details and a host of examples.

In the Lagrangian case the field variables are again given throughout each element, but the values *at* the boundary nodes are used to advance the mesh forward in time. Just as in the Eulerian case, a shape function is required. A function based on one-dimensional cubic splines, guaranteeing continuous first and second derivatives, is desirable. The irregular geometry of the deforming Lagrangian zones requires more thought and creativity than does the Eulerian case. Fracture, sliding interfaces, and changing contacts between neighboring zones are the foci of much current research.

Both finite-element and finite-difference methods have tendencies toward instability unless special measures are employed. Shockwaves can be expected to develop due to the tendency of undamped waves to sharpen as time goes on. The sharpening is caused by the increase of signal speed with density[†], causing the higher-pressure part of the wave to overtake the slower components. Infinitely-sharp shockwaves can be prevented by using an "artificial" (numerical) viscosity large enough to spread shockwaves over a few zones. This additional viscosity makes it possible to extend the length scale of simulation zones by eliminating the large-zone tendency toward turbulence. Unstable Lagrangian modes of deformation with picturesque and descriptive names ("butterfly", "hourglass", and so on) can likewise be prevented through the use of special artificial tensor viscosities. Figure 6.1 shows an example simulation computed both with and without hourglass control.

Figure 6.1: Simulations with (left) and without (right) hourglass control. The initial condition was a periodic sinusoidal density profile with maximum density in the center of the mesh. All the zones have equal masses.

6.6 Smooth Particle Applied Mechanics

A regular spatial grid is not necessary for continuum simulations. Macroscopic continuum simulations can just as well be based on an "unstructured" moving spatial grid made up of "smooth particles". The smooth particles' equations of motion include averaged values of the macroscopic

---

[†]For the usual "polytropic" equation of state $P \propto \rho^\gamma \longrightarrow c = \sqrt{\partial P/\partial \rho} \propto \rho^{(\gamma-1)/2}$.

stress gradient evaluated at each particle's position. It is appealing to solve continuum problems with a particle method. This is because the ordinary differential equations for particle motion are (i) relatively easy to solve, and (ii) free of the instabilities that plague grid-based methods. Smooth-particle methods are based on two ideas: (i) continuum properties—density, velocity, energy, pressure, and heat flux, for instance—have first to be interpolated in space on a discrete but irregular particle grid; (ii) the grid *particles* then exchange energy and momentum according to the *continuum* constitutive relations, where local divergences of the pressure tensor and heat flux vector are evaluated at each particle's location.

In 1977 Monaghan and Lucy independently discovered exactly the same practical scheme for interpolating continuum properties on a moving grid, and for exchanging momentum and energy among particles. The particle locations *define* the moving grid, with spatially-averaged particle properties simultaneously representing the underlying continuum. The spatial influence of each moving particle is described by a normalized short-ranged weight function $w(r)$. *Useful* weight functions need to have at least two continuous derivatives so as to represent solutions of diffusive continuum equations, which incorporate two space derivatives. The simplest example of such a weight function is Lucy's quartic function, shown in Figure 6.2.

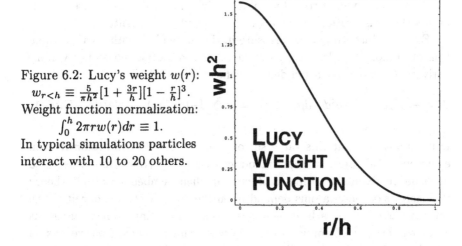

Figure 6.2: Lucy's weight $w(r)$:
$w_{r<h} \equiv \frac{5}{\pi h^2}[1 + \frac{3r}{h}][1 - \frac{r}{h}]^3$.
Weight function normalization:
$\int_0^h 2\pi r w(r) dr \equiv 1$.
In typical simulations particles interact with 10 to 20 others.

At any point in space, r, local values of the mass, momentum, and energy densities are calculated by summing up the contributions of nearby particles. The local mass density $\rho(r) = \rho_r$, for instance, is a superposition of contributions from every particle lying within the range h of the point in question. The same idea can be applied at the particle locations. The mass density at the location of Particle i then follows from a special case of the definition:

$$\rho_r = m \sum_j w_{rj} \longrightarrow \rho_i = m \sum_j w(r_{ij}) = m \sum_j w_{ij}.$$

In such smooth-particle pair sums a typical smooth particle interacts with perhaps twenty of its neighbors. The very smooth character of w, with both ∇w and $\nabla\nabla w$ continuous, guarantees the resulting continuity of both first and second spatial derivatives, such as $\nabla\rho$, $\nabla^2 T$, and $\nabla \cdot P$.

The smooth-particle continuum equation of motion, in the "Lagrangian" form, following the motion, $\rho\ddot{r} = -\nabla \cdot P$, likewise illustrates the simplicity of the gradient operation using smooth particles. The divergence of the pressure tensor, evaluated at the location of Particle i, for instance, becomes a sum of individual particle pressure tensors multiplied by weight-function gradients for all particles within range of Particle i:

$$\{\ddot{r}_i = -m \sum_j [(P/\rho^2)_i + (P/\rho^2)_j] \cdot \nabla_i w_{ij}\}.$$

This particular symmetrized form is a specially nice choice, because it guarantees the conservation of linear (but *not* angular) momentum.

For a fluid isentrope, the pressure in the inviscid smooth-particle equation of motion is hydrostatic, and can therefore be viewed as the volume derivative of a density-dependent specific energy $e(\rho)$:

$$P = -dE/dV = +\rho^2 de/d\rho \longrightarrow \{\ddot{r}_i = m \sum_j [-(de/d\rho)_i - (de/d\rho)_j]\nabla_i w_{ij}\}.$$

It is interesting that this same set of smooth-particle *continuum* motion equations is *identical* to that introduced by Daw, Foiles, and Baskes to describe the dynamics of metal *atoms!*[‡] In their "embedded-atom" theory the energies of metal atoms depend upon the local electronic density. The smooth-particle approach to solving problems in continuum mechanics has a 20-year history of applications to a wide range of flows, as well as to complex deformations in solid mechanics. For references to applications,

[‡]For references see my paper describing the isomorphism, Physica A **260**, 244-254 (1998).

see recent reviews of the method. There is a comprehensive description of the numerical aspects of the method in the Los Alamos report written by Crotzer, Dilts, Knapp, Morris, Swift, and Wingate. Monaghan has been particularly creative in modifying the basic smooth-particle algorithm so as to improve its accuracy and eliminate its instabilities.

Smooth particles exhibit interesting numerical features due to the discretization of space—artificial viscosity, artificial heat conductivity, artificial yield strength, and artificial surface tension. All of these artificial effects vanish for sufficiently small weight function ranges and sufficiently many particles. Two examples illustrate these ideas. Smooth-particle simulation of the shear of an ideal gas, with no transport coefficients whatever, gives rise to an *effective* shear viscosity coefficient and a corresponding turbulent Reynolds' stress. Likewise, smooth-particle simulations of liquid drops can produce oscillations driven by an artificial surface tension. The drop oscillations are described perfectly by Rayleigh's theory. The artificial numerical transport properties can be understood by considering again the smooth-particle equations of motion:

$$\{\ddot{r}_i = -m \sum_j [(P/\rho^2)_i + (P/\rho^2)_j] \cdot \nabla_i w_{ij}\}.$$

Provided that the quotient (P/ρ^2) is slowly varying in space, the equations of motion,

$$\{\ddot{r}_i \propto - \sum_j \nabla_i w_{ij}\},$$

become those of ordinary molecular dynamics, so that the "artificial" viscosities and conductivities are simply Green-Kubo transport coefficients for a fluid with a pair potential proportional to the smooth-particle weight function $w(r)$.

6.7 Example Problems

Here we consider the Rayleigh-Bénard convective flow problem of Section 4.9 from the continuum point of view. The first solution method illustrated is straightforward finite differences, in which horizontal and vertical spatial derivatives are replaced by equivalent differences. To ensure numerical stability it is necessary to define two different spatial grids, one for energy and velocity and another for density, stress, and heat flux. Although this complication is not explicitly included in molecular dynamics, it is certainly reminiscent of the usual convention in molecular dynamics in which the contributions of particle pairs to the pressure and heat flux are thought of as shared *between* the two members of each pair.

I also illustrate the smoothed-particle method—often abbreviated, as "sph" for "smooth-particle hydrodynamics", or "SPAM" for "Smooth Particle Applied Mechanics"—for solving the same problem. In this approach the particles themselves define a (moving and deforming) grid, upon which and within which the continuum properties are defined as weighted sums of particle properties. Although the instabilities encountered with finite-difference or finite-element methods are avoided by using smooth particles, the method requires creativity to adapt it to more-complex boundary conditions, especially those with free surfaces or material interfaces.

6.7.1　*Rayleigh-Bénard Flow with Finite Differences*

The simplest numerical methods are based on solving the conservation laws of continuum mechanics on a fixed Eulerian grid. With gravity added, the corresponding field equations describe the evolving flows of mass, momentum, and energy:

$$\partial \rho / \partial t = -\nabla \cdot (\rho v);$$

$$\partial v / \partial t = -v \cdot \nabla v + g - (1/\rho)[\nabla \cdot P];$$

$$\partial e / \partial t = -v \cdot \nabla e - (1/\rho)[\nabla v : P + \nabla \cdot Q].$$

In numerical work it is preferable to solve equations for the *conserved* variables' densities $\{\rho, \rho v, \rho[e + \frac{1}{2}v^2]\}$, following the derivation of Section 4.1:

$$\partial \rho / \partial t = -\nabla \cdot (\rho v);$$

$$\partial (\rho v) / \partial t = -\nabla \cdot (\rho v v + P) + \rho g;$$

$$\partial (\rho[e + \tfrac{v^2}{2} + gh]) / \partial t = -\nabla \cdot (P \cdot v + \rho v[e + \tfrac{v^2}{2} + gh] + Q) + \rho g v_h,$$

where v_h is the velocity component in the field direction. By using the continuity equation to simplify the last of these equations, it is easy to see that *all* the terms involving the gravitational acceleration g cancel, leaving *exactly the same* form of the energy equation as in the field-free case:

$$\partial(\rho[e + \tfrac{v^2}{2}])/\partial t = -(\nabla v : P) - \nabla \cdot (\rho v[e + \tfrac{v^2}{2}] + Q).$$

The three coupled equations for $\{\rho, \rho v, \rho[e + \tfrac{1}{2}v^2]\}$ need to be solved everywhere within the system's boundaries. For this approach to be well-posed, these equations need to be "closed" with additional constitutive relations, giving the functional dependence of the pressure tensor P and the heat flux vector Q on the field variables. Initial conditions for the density, velocity, and energy also need to be provided, as well as *thermal* boundary conditions specifying the temperature or the heat flux and *mechanical* boundary conditions specifying the velocity or the pressure loads.

The motionless boundaries of the Rayleigh-Bénard problem, hot on the bottom and cold on the top, make this the simplest of nonequilibrium flows. See the Figures which follow for the flow geometry. For simplicity I consider a two-dimensional $L \times 2L$ system, with *periodic* boundaries at the sides. An alternative solution for this same problem, using molecular dynamics, was discussed in Section 4.9.2. The only addition to conventional molecular dynamics required there was a thermostating mechanism incorporated into the upper and lower thermal boundaries.

Here I consider the continuum analog. The Eulerian equations solved in the numerical work are mathematically equivalent to the simpler Lagrangian *partial* differential equations:

$$\dot{\rho} = -\rho \nabla \cdot v \; ; \; \rho \dot{v} = \nabla \cdot \sigma + \rho g \; ; \; \rho \dot{e} = \sigma : \nabla v - \nabla \cdot Q,$$

but the Eulerian form, which evaluates changes in mass, momentum, and energy as the result of fluxes into, and out of, each element, makes it possible to conserve mass, momentum, and energy *exactly*. The solution must satisfy the *boundary conditions* also, with vanishing velocity and fixed temperature at the hot and cold boundaries.

The simplest method for solving this problem uses two grids, within which values of all the field variables $\{\rho, v, e, \sigma, Q, \dots\}$ can be linearly interpolated as needed. For stability the velocities and energies are specified on one grid and the density, stress, and heat flux on the other. For convenience the velocity-energy grid coincides with the system boundary. By using the mechanical and thermal equations of state for a two-dimensional monatomic ideal gas: $PV = E = Nme = NkT$, the thermal boundary conditions give fixed values for e, the internal energy per unit mass, at all the top and bottom boundary nodes. The velocity components at these same nodes are all set equal to zero.

Histories of density, as well as those of the pressure tensor and the heat flux vector, are followed at the *centers* of the grid cells. The gradients are calculated as needed using centered spatial differences or the more-nearly-accurate derivatives of cubic splines. To relate the velocity and temperature gradients to stress and heat flux, we use the simplest possible assumption, constant transport coefficients.

Sample results for three Rayleigh numbers are given in the Figures. This work is described in detail in Vic Castillo's 1999 PhD thesis (University of California at Davis/Livermore) and two of his 1997 papers in Physical Review E. Within particular ranges of the temperature differences and transport coefficients, two or more different stable solutions can sometimes be found. Which one is observed can then be sensitive to the initial conditions in a complicated way. Thus this simplest example problem illustrates the lack of uniqueness in "stationary" solutions of the continuum equations. This bifurcation also implies that sufficiently-large-scale molecular dynamics simulations *cannot* always be ergodic when applied to flows as complex as is the "simple" Rayleigh-Bénard problem.

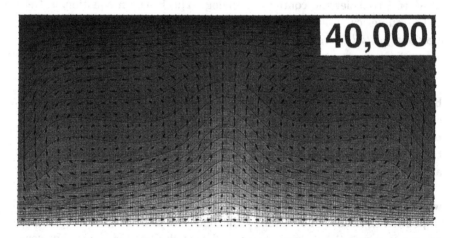

Figure 6.3: Instantaneous flow velocities and corresponding temperature contours in a stationary Rayleigh-Bénard flow. The Rayleigh Number is 40,000. The fluid is an ideal gas with constant transport coefficients.

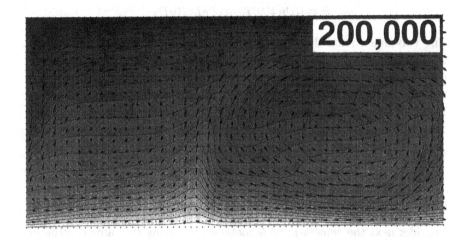

Figure 6.4: Instantaneous flow velocities and temperature contours in an oscillating Rayleigh-Bénard flow. The Rayleigh Number is 200,000. The fluid is an ideal gas with constant viscosity and heat conductivity.

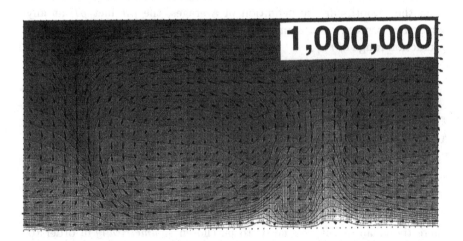

Figure 6.5: Flow velocities and temperature contours in a chaotic Rayleigh-Bénard flow with thermal plumes. The Rayleigh Number is 1,000,000. The fluid is an ideal gas with constant viscosity and heat conductivity.

As the temperature difference is increased, the Rayleigh Number rises:

$$R \equiv (\partial \ln V / \partial T)_P \Delta T g L^3 / (\nu D_T).$$

The two-dimensional flow then passes reproducibly through several distinct stages: (i) stationary conduction; (ii) steady convection, with roughly-circular rolls transporting heat upward; (iii) oscillating vertical motion of the rolls; (iv) formation of convecting plumes. In this last stage, the flow can become very irregular, and can exhibit transitions between two branches, one which oscillates regularly and the other which behaves more chaotically. Then the two branches (iii) and (iv) are separated by a hysteresis loop.

6.7.2 *Rayleigh-Bénard Flow with Smooth Particles*

Consider next the *smooth-particle* method for solving the same continuum flow problem. The smooth-particle method defines the local densities at the particles as sums of particle contributions: $\{\rho_i = \sum_j m w_{ij}\}$. The weight function $w(r)$ gives the influence of a particle at a distance r. The gradients of local particle properties computed by superposition, follow by straightforward differentiation. Consider, for instance, the temperature gradient. From the smooth-particle definition of averages, the product ρT at any location r is given by a sum over all particles $\{j\}$ which lie within the interaction range of r:

$$(\rho T)_r \equiv \sum_j m T_j w(r - r_j).$$

The spatial gradient of ρT is given by: $(\rho \nabla T)_r + (T \nabla \rho)_r = \sum_j m T_j \nabla_r w_{rj}$. An evaluation of this gradient at the location $r \equiv r_i$ of a particular particle, Particle i, can be solved for the local temperature gradient there, $(\nabla T)_i$:

$$(\nabla T)_i = \sum_j m[(T_j - T_i)/\rho_{ij}] \nabla_i w_{ij},$$

where the mean density ρ_{ij} needs to be chosen *symmetrically* in order to guarantee the conservation of energy:

$$\rho_{ij} \equiv (\rho_i + \rho_j)/2 \text{ or } \rho_{ij} \equiv (\rho_i \rho_j)^{1/2}.$$

With all the individual temperature gradients known the smooth-particle form of the energy equation evaluated at the particle locations gives:

$$\{\dot{e}_i = -\sum_j (m/2)[(P/\rho^2)_i + (P/\rho^2)_j] : (v_j - v_i)\nabla_i w_{ij}$$

$$-\sum_j m[(Q/\rho^2)_i + (Q/\rho^2)_j] \cdot \nabla_i w_{ij}\} \ ; \ \{Q_i = -\kappa(\nabla T)_i\}.$$

These equations, together with the corresponding equations of motion:

$$\{\dot{r}_i = v_i \ ; \ \dot{v}_i = g - \sum_j m[(P/\rho^2)_i + (P/\rho^2)_j] \cdot \nabla_i w_{ij}\},$$

where g is the gravitational acceleration, give a set of $5N$ ordinary differential equations for two-dimensional problems $\{\dot{x}, \dot{y}, \dot{v}_x, \dot{v}_y, \dot{e}\}$ and $7N$ equations in three dimensions $\{\dot{x}, \dot{y}, \dot{z}, \dot{v}_x, \dot{v}_y, \dot{v}_z, \dot{e}\}$.

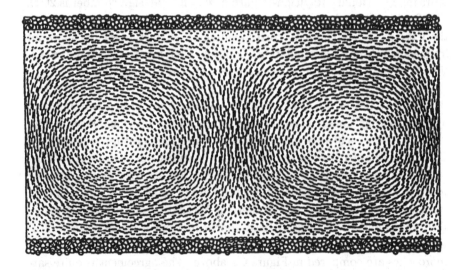

Figure 6.6: Reflected image particles, indicated by circles, follow the motion of corresponding bulk particles. The 5000 bulk particles shown here simulate the Rayleigh-Bénard flow of an ideal gas with constant transport coefficients at a Rayleigh Number of 10000. For details see the paper by Kum, Hoover, and Posch.

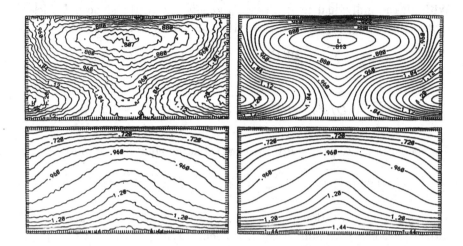

Figure 6.7: Comparison of density (above) and temperature (below) contours for a stationary Rayleigh-Bénard flow. The Rayleigh Number is 2000. The fluid is an ideal gas with constant viscosity and heat conductivity. At the left are shown contours corresponding to an instantaneous snapshot from a smooth-particle simulation. Contours from a corresponding conventional Eulerian finite-difference simulation appear at the right.

Boundary values of the velocity and temperature are most easily imposed by constructing "image particles", as shown on the previous page, in Figure 6.6, with specified boundary values of the temperature and velocity. The complete set of ordinary differential equations can then be solved using the Runge-Kutta method described in Section 5.2. Deviations from the accurate extrapolated continuum solution are of order $N^{-1/2}$, as would be expected from the magnitude of thermal fluctuations in particle systems. The typical deviations of grid-based approximations from an accurate Eulerian solution are generally considerably smaller. Results from the two approaches are compared in Figure 6.7 above. The agreement is quite satisfactory.

6.8 Summary

Simulating macroscopic flows presents additional difficulties and instabilities not present in atomistic simulations. Spatial discretization introduces special possibilities for numerical difficulties in Eulerian simulations which follow interfaces and in Lagrangian simulations which seek to follow the flow. These difficulties are usually overcome by introducing *ad hoc* viscosities to slow or eliminate the instabilities. The errors due to discretization can introduce paradoxical transport and constitutive properties leading to artificial dissipation and structure.

Because macroscopic simulations involve more complex properties (stress and heat flux, rather than just coordinates and velocities) and also require interpolation to find fluxes and derivatives, the numerical work is increased by roughly an order of magnitude. Smooth-particle methods provide an interesting compromise, allowing continuum constitutive relations to dictate the motion of an interpenetrating set of representative particles. Stability is improved, but accuracy is hindered by the enhanced fluctuations intrinsic to particle methods as well as by the difficulties in locating interfaces and preventing interpenetration.

Chapter 7

Chaos, Lyapunov Instability, Fractals

The brain is wider than the sky,
for put them side by side.
The one the other will include
with ease, and you beside.

Emily Dickinson

7.1 Introduction

The mathematical ideas necessary to a quantitative description of chaos languished during the years that quantum mechanics attracted the attention of most physicists. Poincaré, Lyapunov, and Krylov had already discussed and detailed the exponential growth of small perturbations. Cantor had described and displayed fractal sets exhibiting fractional dimensionality. These two elements—exponential growth and fractals—are the common features of time-reversible dissipative flows. With time symmetry present the equations of motion guarantee that any system exhibiting phase-space shrinkage (to its attractor) must also have a similar, but concealed, region of unstable phase-space growth (from its repellor). The complete structure can be thought of as a "simple" flow, *from* the repellor source *to* the attractor sink. The simplicity is actually more than a little illusory because both the attractor and repellor sets are typically *ergodic*, dense everywhere. As this flow passes from the expanding region near the repellor to the shrinking region near the attractor, the sum of local Lyapunov exponents shifts from positive to negative.

When it became possible to use computers to generate and visualize solutions of coupled nonlinear equations, widespread realization dawned that
unpredictability, based on sensitivity to small perturbations, was characteristic of such problems. Lorenz' Attractor, published in 1963 and discussed
in Section 4.9.1, soon became a widely-known example of unpredictability.
Lorenz documented the sensitive dependence on the initial conditions displayed by his simple model for Rayleigh-Bénard convection. This model is
the set of three coupled quadratic equations:

$$\{\dot{x} = -\sigma(x - y) \; ; \; \dot{y} = \mathcal{R}x - y - xz \; ; \; \dot{z} = xy - bz\}.$$

He concluded that nature could hardly be less complex than this "simple"
system, making the prediction of weather problematic. He showed that successive maxima of his variable z (proportional to the vertical heat transfer
rate) have nearly the same form as iterates of the irreversible "tent map",

$$z'(0 < z < 1) = 1 - |1 - 2z|,$$

which, like the Baker Maps, creates and destroys information by shifting
less-significant bits to more-significant positions, from which they are ultimately discarded.

Weather has no monopoly on the instabilities we call "chaos" or "Lyapunov instability". *Whenever* pervasive small-scale microscopic divergences
are localized, on a larger macroscopic scale, chaos is the result. The growth
of very small chaotic perturbations is exponential—Lyapunov unstable.
The localization of the chaotic motion imposed by geometric or energetic
constraints, then leads to the complex structures familiar from computer
displays. I avoid here a common alternative phrase for chaos, "sensitive dependence on initial conditions", because the initial conditions in computer
simulations are never precisely sharp. They are limited by the number of
digits carried. Further, the numerical methods which advance the time
continually inject additional "information" or "noise" or "errors" as the
calculation proceeds. We are certainly free to imagine the hypothetical existence of an idealized "exact solution" of the motion equations. But such
a solution has no real existence, even in principle—any feasible approximate numerical "solution" is continually perturbed away from it. Joseph
Ford repeatedly emphasized, as had Maxwell a century earlier, that there
is no way, even in principle, to construct a (precise) solution of the classical
motion equations in the presence of deterministic chaos. Such solutions

are intrinsically unknown and unpredictable. For any fixed level of uncertainty, the number of digits kept, in the initial conditions and in the computational algorithm, both increase in proportion to the total time for which an approximate numerical "solution" is required.

For applications to reproducible real-world problems the lack of a "true" infinitely-precise solution is unimportant. A good approximation suffices. The computational situation resembles that facing experimentalists, for whom it is commonplace that the irrelevant experimental details are neither interesting nor reproducible. The successful experimentalist, like his computational counterpart, selects those features common to a wide body of potential experiments for compilation, analysis, and understanding.

From a more theoretical outlook the instabilities which amplify small perturbations are extremely interesting. Our current understanding is based on dynamical and topological ideas combined with the computers and the algorithms necessary to try them out. Simple topological analyses of Lyapunov instability follow the distortions of a small comoving "ball" or hypersphere of neighboring solutions. For short times the moving fine-grained ball becomes a rotating hyperellipsoid, with well-defined principal axes. The relative rates of growth and decay parallel to these rotating axes define the local "Lyapunov Spectrum", which is intimately related to the macroscopic dissipation through the *instantaneous* sum rule:

$$\dot{\otimes}/\otimes \equiv \sum \lambda_{\mathrm{L}}.$$

As usual $\otimes$ and $\dot{\otimes}$ denote a small comoving volume element, and its time derivative following the motion, in the phase space. In order that the local exponents be well-defined, the comoving corotating vectors defining them—see Section 7.4—need to have been followed forward in time from an initial trajectory point sufficiently far in the past. Though the local exponents $\{\lambda_{\mathrm{L}}\}$ *do* depend upon the chosen coordinate system, their long-time averages, *the* Lyapunov Spectrum $\{\lambda \equiv \langle \lambda_{\mathrm{L}} \rangle\}$ do not. It is typical of the local exponents that their *time* dependence is regular while their *spatial* variation is singular.

In "dissipative" systems the summed spectrum of the (time-averaged) Lyapunov exponents is negative, signaling an overall decreasing phase volume $\langle \dot{\otimes} \rangle < 0$. Obviously, the long-time limit of that comoving fine-grained volume must vanish *despite* the exponential divergence of small perturbations! Lorenz' system of three differential equations is among the simplest of the continuous systems that show this behavior. For a wide range of parameter values Lorenz' system produces chaos, with exponential growth of perturbations in a confined region of (x, y, z) space. His system, which includes the decays:

$$\{\dot{x} \propto -x \; ; \; \dot{y} \propto -y \; ; \; \dot{z} \propto -z\},$$

is obviously not time-reversible. An equally-simple set of three differential equations, which *is* time-reversible, and which also shows chaos for some initial conditions, but not for others, is the "Nosé-Hoover oscillator":

$$\{\dot{q} = p/m \; ; \; \dot{p} = -\kappa q - \zeta p \; ; \; \dot{\zeta} = [(p^2/mkT) - 1]/\tau^2\}.$$

The control variable, or "friction coefficient" ζ maintains the long-time average temperature, $T = \langle p^2 \rangle / mk$. The time-reversed trajectory is constructed by the transformation:

$$(+q, +p, +\zeta) \longrightarrow (+q, -p, -\zeta),$$

coupled with a reversal of the time ordering of the points. The chaos of the Nosé-Hoover oscillator is not "confined" in the usual sense because the stationary density distribution,

$$f(q, p, \zeta)_{\text{eq}} \propto e^{-\kappa q^2/2kT} e^{-p^2/2mkT} e^{-\zeta^2 \tau^2/2},$$

extends over *all* values of the three dependent variables. This regular distribution gives way to a complex multifractal if the temperature T varies with q.

For systems with a physical interpretation, such as Lorenz' dissipative model for Rayleigh-Bénard convection and the Nosé-Hoover oscillator with a variable temperature, strange attractors have physical significance. They indicate first of all the rarity of the represented chaotic states. In most cases the rate of collapse of their phase volume corresponds also to the external dissipation rate $\dot{S}/k$ and to the information loss rate. Both features of these strange attractors are common to large classes of physical problems. External dissipation and an exponentially fast collapse of the extension in phase $\otimes$ both correspond to macroscopic irreversible behavior, as described by the Second Law of Thermodynamics.

In this Chapter I describe the tools necessary to take the measure of chaos: the Lyapunov spectrum and the fractal dimensionalities which characterize nonequilibrium phase-space structures. Let us begin by describing the structure of the stage upon which this drama unfolds.

7.2 Continuum Mathematics

The *discontinuous* nature of computation might seem unsuited to the exploration of fine-grained fractals generated by *continuous* equations. But a little probing reveals this apparent problem to be an insubstantial straw man. There is no compelling reason, other than the pursuit of simplicity, to distinguish between the analytic and numerical descriptions of physical systems. The *analytic* work is simplest for the purely-hypothetical infinitely-precise continuous description of coordinates, velocities, and time. Some of the blind alleys constructed with this description by mathematicians searching for Cantor's paradise were illuminated and mapped by Bridgman in 1934. *Numerical* work *requires* a *discretized* description of all these variables. Computation is simplest for a fixed (finite) word length corresponding to about fourteen decimal digits. From an operational standpoint the fully continuous case can consistently be regarded as a limiting case of the natural computational representation of space and time by rational numbers. The movable scalable finite grid of computer numbers contains equal fractions of binary 0's and 1's, and has exactly the same *fractal dimensionalities* (see Section 7.5), both box-counting and information, as does the underlying hypothetical continuum. Wherever one looks, and however finely, there are fourteen orders of magnitude with continuum properties before the level of neglect of 10^{-14} is reached. Any admixture of 0's and 1's *other* than equal fractions corresponds to a set with a family resemblance to Cantor's, with a singular distribution and a fractional information dimension, relative to the continuum.

Numerical solutions of differential equations naturally occupy the nodes of a computational lattice, while the probability density, or measure, generated by accumulating a long trajectory is often more naturally viewed as a continuous function. Probability density can be thought of as coarse-grained, averaged over (infinitesimal) regions "between" adjacent nodes of the computational lattice. It is interesting that coarse-graining was a familiar idea to Boltzmann and Gibbs half a century before the construction of the first computers. It is just as *natural* to measure and analyze the fractal distributions generated by computers with arrays of boxes as it was *natural* to discover and apply integral calculus.

7.3 Chaos

Systems with a "chaotic" time dependence display a disorderly lack of regularity and predictability, due to the pervasive global exponential growth of small perturbations and the absence of periodicity. Isolated local exponential behavior is not enough for chaos. The inverted rigid pendulum, for instance, is perfectly predictable and periodic. Its motion is not "chaotic". *Two* coupled rigid pendula, one supporting the other, *are* enough for chaos. Two coupled pendula behave unpredictably whenever they lie within the chaotic regions of their four-dimensional phase space. The corresponding system trajectory develops in a three-dimensional constant-energy subspace. If the trajectory were ever exactly to intersect itself the motion would in theory repeat. But the inevitable uncertainty in any trajectory, either physical or numerical, is amplified so as to avoid this repetition. Evidently the distinction between a trajectory which "eventually" intersects itself and one which "never" does is at best an ill-posed problem for the philosophers, as it lies beyond computation.

Maxwell, Boltzmann, and Poincaré had a crude knowledge of what is now called "chaos". The knowledge was of little use until it could be implemented on the fast computers which became available at the close of the Second World War. Lorenz' well-known butterfly-shaped attractor captured the interest and imagination of computational scientists. Consider again Lorenz' set of three differential equations in the variables (x, y, z):

$$\{\dot{x} = -\sigma(x - y) \; ; \; \dot{y} = \mathcal{R}x - y - xz \; ; \; \dot{z} = xy - bz\}.$$

He discovered that perturbations $\{(\delta x, \delta y, \delta z)\}$ typically grew exponentially in the time, exhibiting "Lyapunov instability". Nevertheless, despite the exponential growth the motion is confined to a fractional-dimensional region. An example, for $\{\sigma = 10, \mathcal{R} = 28, b = 8/3\}$, appears on page 105. As a result of Lorenz' work a new kind of mathematical object, a "strange attractor", became familiar to physicists. The name emphasizes the *simultaneous* coexistence of exponential divergence with a fine-grained *contracting* volume which has a *vanishing* limit. For *interesting* strange attractors the traditional perturbation expansions about *fixed points*, classified as elliptical, parabolic, or hyperbolic, are of little use. Furthermore, many sets of motion equations, like those of the Nosé-Hoover oscillator, for which we make the simplest choice, unity, for the parameters (κ, m, τ, kT):

$$\{\dot{q} = p \; ; \; \dot{p} = -q - \zeta p \; ; \; \dot{\zeta} = p^2 - 1\},$$

have *no fixed points at all*.

Purely Hamiltonian chaos has attracted considerable attention to a variety of physics problems over a wide range of space scales. Small-scale particle accelerators and large-scale astrophysical systems can both exhibit complex structures incorporating stable as well as unstable regions in their solution spaces. Although fascinating, such intricate structures are not useful for validating or understanding statistical mechanics, where it is fervently desired that any interesting *macroscopic* results *not* vary with the *microscopic* initial conditions. Though general proofs establishing the ergodicity property understandably elude the efforts of mathematicians, by now there are *many* small nonequilibrium systems for which computation has demonstrated ergodicity. This is a *typical* feature of long-time-averaged nonequilibrium steady states. And provided that *any* initial condition generates the *same* strange attractor, the time-reversibility of the equations of motion implies ergodicity, because any trajectory point lying in the past corresponds also to a possible trajectory "image point"—a point with all the velocities reversed—which will necessarily be encountered in the future. Because exhaustive sampling is the only way to check ergodicity, systems chosen for investigation have to have only a few degrees of freedom. For the simplest of systems, some theoretical analysis is possible—Sinai is credited with proving the quasiergodicity of hard-disk and hard-sphere systems at equilibrium. In either the conservative or the dissipative case, chaos can be quantified through the spectrum of Lyapunov exponents, to which we turn next.

7.4 Spectrum of Lyapunov Exponents

Let us consider stationary ergodic flows. Exponential instabilities for these flows are described by the Lyapunov exponents $\{\lambda\}$, as indicated in Figure 5.2 of Section 5.5. These "Lyapunov exponents" are long-time averages of "local" exponents $\{\lambda_L\}$, giving the orthogonal growth and decay rates of corotating basis vectors, one for each independent direction in the embedding space. Conventionally the exponent with the largest long-time-averaged value is λ_1, that with the second-largest average is λ_2, and so on. One of the exponents, which corresponds to the growth rate along the direction of motion in the phase space, has a time-averaged value of zero. Because the *ordering* of the local exponents changes (the local exponents likewise typically change sign) the numbering convention, with $\lambda_i > \lambda_{i+1}$, requires *global* information. For short times, the local exponents depend upon the initial conditions. The Lorenz model has three such exponents. Though their sum is fixed,

$$\dot{\otimes}/\otimes = \sum \lambda_{\mathrm{L}} = -\sigma - 1 - b,$$

the three individual local exponents fluctuate wildly, but smoothly, with time, and show all possible combinations of the signs of λ_1 and λ_2, but with λ_3 consistently negative. The smooth time dependence contrasts with their highly-singular spatial dependence. See Figure 7.1.

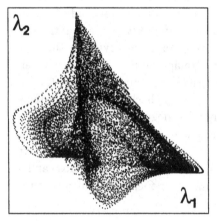 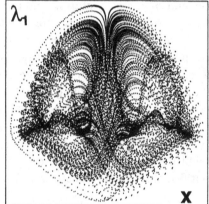

Figure 7.1: Distribution of the local Lorenz attractor Lyapunov exponents. The attractor parameters are $\{\sigma = 10, \mathcal{R} = 28, b = 8/3\}$. The left view shows the correlation $\{-10 < \lambda_1 < +10, \ -6 < \lambda_2 < +8\}$ linking the *local* exponents, λ_1 and λ_2. The right view shows the variation of the *local* value of the largest exponent $-10 < \lambda_1 < +10$ with Lorenz' variable $-20 < x < +20$. The sum of all three local exponents is a constant of the motion, $-13\frac{2}{3}$.

Even for conservative dynamical systems the individual *local* values depend upon the chosen coordinate system, while the long-time averages do not. Thus the local exponents do not really characterize the underlying physical system. To avoid clutter I will often use the same notation $\{\lambda\}$ for both the local and the time-averaged Lyapunov exponents, trusting to the reader to distinguish the two concepts by context.

Let us consider the calculation of the complete Lyapunov spectrum for a general dynamical system, with motion equations $\dot{x} = \mathcal{F}(x)$. The time-dependent variables x define a vector locating the system in its embedding space. Begin with a "reference trajectory", a solution of the equations starting from a definite point, but sufficiently long ago that the identity of this initial point is unimportant. In the neighborhood of this solution of the motion equations, the equations of motion can be linearized, so that the vector separation, from the reference trajectory to any nearby "satellite trajectory" solution, is the infinitesimal vector $\delta(t)$, which has a motion equation determined by the first derivative of the reference system's motion equations:

$$\dot{\delta} \equiv D \cdot \delta \; ; \; D \equiv \partial \mathcal{F} / \partial x.$$

For clarity, let us consider a simple example of this general approach. For the Nosé-Hoover oscillator of the last section, governed by the equations,

$$\mathcal{F} = \{\dot{q} = p \; ; \; \dot{p} = -q - \zeta p \; ; \; \dot{\zeta} = p^2 - 1\},$$

the *relative* motion of a satellite trajectory, with "offset vector",

$$\delta \equiv (\delta q, \delta p, \delta \zeta) \equiv (q, p, \zeta)_{\text{satellite}} - (q, p, \zeta)_{\text{reference}},$$

follows the *linearized* motion equations:

$$\{\dot{\delta q} = \delta p \; ; \; \dot{\delta p} = -\delta q - \zeta \delta p - p \delta \zeta \; ; \; \dot{\delta \zeta} = 2p \delta p\}.$$

Notice that these motion equations are time-reversible provided that the original flow was. Thus the reversed evolution of a corotating and comoving hyperellipsoid, following the flow forward in time, generates an image of the repellor, with the signs of all the local Lyapunov exponents reversed $\{+\lambda\} \longrightarrow \{-\lambda\}$. Because the reversed trajectory is actually unstable, these hypothetical precisely-reversible repellor exponents $\{-\lambda\}$ are not observable.

For our Nosé-Hoover oscillator example the "dynamical matrix" $D = \partial \mathcal{F} / \partial x$ governing the evolution of the offset vector δ is:

$$D = \begin{matrix} 0 & 1 & 0 \\ -1 & -\zeta & -p \\ 0 & 2p & 0 \end{matrix}.$$

The maximum Lyapunov exponent is the (time-averaged) rate of growth of the unconstrained infinitesimal vector δ and describes the rate of divergence of a nearby satellite trajectory away from the reference trajectory. Spotswood Stoddard and Joseph Ford used this idea to compute the largest Lyapunov exponent for a dense Lennard-Jones fluid in 1973. The second-largest Lyapunov exponent, λ_2, is defined in the same way, but for an offset vector δ_2 which is required to be *orthogonal* to δ_1. This idea generalizes to the computation of the entire spectrum*. The exponents have an interesting topological interpretation: the sum of the first n time-averaged exponents gives the global growth rate of an $(n+1)$-point n-dimensional object in the embedding space. We discuss the consequences of this observation in the following Section.

In following the offset vectors $\{\delta\}$ linking the reference trajectory to the satellites (one for each exponent) forward in time it is absolutely necessary to rescale them, either continuously, or at fixed time intervals. Otherwise their size will rapidly grow or shrink beyond the possible limits of computational precision. This rescaling operation is analogous to the velocity rescaling used in isokinetic thermostats. The orthogonality of the vectors also needs to be explicitly maintained.

The evolution of the local Lyapunov exponents can then be followed by solving a coupled set of equations, with the orthogonality and rescaling restrictions included, using a separate offset vector δ for each Lyapunov exponent. If the original dynamical system has N independent variables, then a system of $N(N+1)$ equations needs to be solved for the N variables defining the reference trajectory and the N^2 offset-vector components of the N satellite trajectories. In the Nosé-Hoover oscillator example, the complete spectrum of three exponents requires the solution of twelve ordinary differential equations, three for the reference trajectory and nine for the three orthogonal satellite trajectories. If N is not too large an elegant

---

*Giancarlo Benettin developed a numerical algorithm for the spectrum in 1976.

Lagrange-multiplier approach[†] can be used to impose the N length constraints $\{\delta_i^2 \equiv 1\}$ and the $N(N-1)/2$ orthogonality conditions $\{\delta_i \cdot \delta_j \equiv 0\}$. The general idea is clear enough from the three-exponent example:

$$\dot{\delta}_1 = D \cdot \delta_1 - \lambda_{11}\delta_1;$$

$$\dot{\delta}_2 = D \cdot \delta_2 - \lambda_{21}\delta_1 - \lambda_{22}\delta_2;$$

$$\dot{\delta}_3 = D \cdot \delta_3 - \lambda_{31}\delta_1 - \lambda_{32}\delta_2 - \lambda_{33}\delta_3.$$

For convenience, let the constant infinitesimal scalar lengths of all the offset vectors be ϵ. Then the *diagonal* Lagrange multipliers, $\{\lambda_{ii}\}$, which exactly compensate for the stretching and shrinking of the unconstrained vectors, can be calculated from the N constant-length conditions:

$$\{(d/dt)\delta_i^2 = 0 = 2\delta_i\dot{\delta}_i \longrightarrow \lambda_{ii} = (\delta_i/\epsilon) \cdot D \cdot (\delta_i/\epsilon)\}.$$

The *off*-diagonal Lagrange multipliers are chosen to keep the offset vectors $\{\delta_1, \delta_2, \delta_3, \dots\}$ orthogonal. These off-diagonal multipliers follow from the $N(N-1)/2$ conditions:

$$\{(d/dt)\delta_i\delta_j = 0 = \delta_i\dot{\delta}_j + \delta_j\dot{\delta}_i \longrightarrow \lambda_{ij} = (\delta_i/\epsilon) \cdot D \cdot (\delta_j/\epsilon) + (\delta_j/\epsilon) \cdot D \cdot (\delta_i/\epsilon)\}.$$

The Lyapunov exponents are themselves the long-time averages of the *diagonal* Lagrange Multipliers:

$$\lambda_1 = \langle\lambda_{11}\rangle \; ; \; \lambda_2 = \langle\lambda_{22}\rangle \; ; \; \lambda_3 = \langle\lambda_{33}\rangle.$$

The spectrum of exponents can, in some exceptional circumstances, exhibit interesting symmetry properties. In Hamiltonian mechanics, the reversibility of the equations of motion, together with the lack of dissipation, indicates that any expanding direction is converted to a contracting direction in the time-reversed flow. This reversed flow and the forward flow are both equally stable. Thus Hamiltonian Lyapunov spectra are made up of *pairs* of exponents $\{+\lambda, -\lambda\}$. If all the momenta are similarly thermostated, with the same friction coefficient, the result can be to shift each pair of exponents by the same amount, $-\zeta$, a rule called "conjugate pairing" by its discoverers, Denis Evans and Gary Morriss.

[†]Discovered by Harald Posch and me in 1986 and rediscovered many times since.

Figure 7.2 illustrates typical equilibrium spectra for fluids and solids in two and three space dimensions. The *fluid* spectra look much like the Debye vibrational spectra of *solid*-state physics. Although the matrix D is common to both approaches (Lyapunov spectra and vibrational frequencies) the exact connection between them is just now becoming understood. Work by Harald Posch and Christoph Dellago has established that the individual mode components $\{\delta q, \delta p\}$ oscillate with a *common phase* in the long-wavelength Lyapunov eigenvectors. This is quite different to long-wavelength sound waves, for which δq precedes δp by a phase shift of $\pi/2$.

Figure 7.2: Typical many-body Lyapunov spectra for both two-dimensional and three-dimensional fluids and solids. The positive half of the spectrum is shown here. See the 1989 paper by Posch and me.

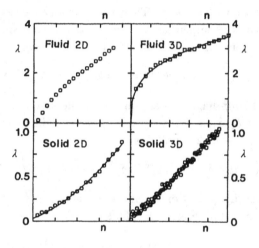

From the theoretical standpoint of kinetic theory, hard spheres are somewhat simpler to treat than are soft particles. Simulations with hard spheres, designed to characterize their Lyapunov spectra, are complicated by the impulsive forces. Nevertheless, recent results from simulations and kinetic theory are beginning to agree. Definitive spectral work for hard disks and spheres in the large-system limit will require modest improvements in computer speeds. The most recent work has just been reviewed by Harald Posch and Robin Hirschl. Their review should appear soon in the *Encyclopædia of Mathematical Sciences*.

7.5 Fractal Dimensions

The two- or three-dimensional macroscopic mass, momentum, and energy densities from continuum mechanics, as well as the many-dimensional microscopic probability densities from Gibbs' equilibrium statistical mechanics, are typically smooth and differentiable, with the same dimensionalities as the embedding dimensions $\{D_E\}$ of the spaces in which they are embedded. Away from equilibrium *probability densities* behave in a different way—they are typically multifractals, with *fractional* dimensionality. These reduced fractal dimensions are most easily defined by generalizing the usual notion of one-, two-, and three-dimensional objects to include objects whose dimensionality is not an integer. Suppose that a sufficiently compact geometric "object" with dimensionality D covers an arbitrarily large number # of sufficiently-small cells (or "boxes"), of infinitesimal width ϵ in an embedding space with dimensionality D_E greater than D. The number covered is of order ϵ^{-D} for small enough cells. For example, a straight line piercing a cube composed of 10^6 small cubes, intersects a number of these smaller cubes of order 100. A plane piercing the same large cube intersects on the order of 10^4 of the small cubes. Thus the ("box-counting") dimension of ordinary one-, two-, and three-dimensional objects can be found by taking the limiting small-box ratio:

$$\ln(\#)/\ln(1/\epsilon) \longrightarrow \ln(\epsilon^{-D})/\ln(\epsilon^{-1}) \equiv D_{\text{BC}},$$

where # is the number of infinitesimal cubes intersecting the embedded object. This definition can be generalized to the nonintegral case; the resulting *box-counting* dimension is sometimes termed the "Hausdorff dimension". Here I avoid that latter terminology, so as to avoid confusion with more interesting and useful ideas, the "information dimension" and the "Kaplan-Yorke dimension" defined below. For the attractors representing typical time-reversible nonequilibrium systems the information dimension *is* sensitive to dissipation while the box-counting dimension is not.

Fractals are distributions of points in which the density of nearby points varies as a power law in the vicinity of each point. If the power law itself varies from point to point the distribution is termed "multifractal". Most textbook examples of fractals contain holes, while those representing physical systems on digital computers appear instead to represent continuous, and often *ergodic*, multifractals. The simplest fractal, *with* holes, is probably Cantor's set. This set can be constructed recursively, as is indicated in Figure 7.3. It is made up of all those base-3 numbers which contain only 0's and 2's in their base-3 expansion. To construct a coarse-grained approximation to this set, with N ternary digits, divide the unit interval,

$$0.000000 \ldots = 0 < x < 1 = 0.222222 \ldots ,$$

into 3^N equal bins (or "boxes") of width 3^{-N}. Then discard any number with a "1" in its first N digits. Evidently the entire interval from 1/3 to 2/3 is discarded, and so not included in Cantor's set, for all such numbers begin with the digit 1: 0.1... . Likewise all the numbers from 1/9 to 2/9 and 7/9 to 8/9 are excluded, for they have a 1 in the second position: 0.01... and 0.21... . For *any* set of 3^N similar boxes, the number of *occupied* boxes is only 2^N, so that the box-counting dimension of this Cantor set is:

$$D_{\text{BC}} = \ln(\#)/\ln(1/\epsilon) = (N \ln 2)/(N \ln 3) = 0.63093.$$

The set is "self similar". Enlarged versions of each of the 2^N intervals isomorphic to $0 < x < 3^{-N}$, when scaled up by factors of 3^N, would appear identical to the Cantor's set occupying the entire original unit interval $0 < x < 1$.

Figure 7.3: Development of Cantor's Set by the repeated removal of the middle-third segments of the unit interval.

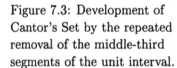 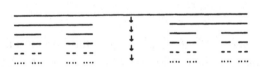

Cantor's Set is a *mathematical object*, not a physical one. The *density* of points in the Cantor set has no special physical significance. By way of contrast, time series of phase-space points, generated by solving the equations of motion for many-body systems, represent equally-likely dynamical states, and can be used to define a variety of fractal dimensions. Though the data represent samples from a hypothetical one-dimensional trajectory in phase space, sufficient points can be generated to characterize the many-dimensional coarse-grained probability density in phase space.

The most useful of the many fractal dimensions characterizing that density is the "information dimension" D_{I}, which gives the limiting (small-box) dependence of the box probability p on the box size ϵ. If the probabilities vary as the Dth power of the box size, $p \simeq (\epsilon^D)$, then the averaged variation defines the information dimension as follows:

$$D_I = \langle \ln p \rangle / \ln(\epsilon) \equiv \sum p \ln p / \ln(\epsilon) \longrightarrow \langle \ln(\epsilon^D) \rangle / \ln(\epsilon) = D_I.$$

The box probability p is the product of the box volume, $\propto \epsilon^{D_E}$—where D_E is the embedding dimension—and the probability density f. As a fringe benefit, this information dimension gives explicitly the dependence of the coarse-grained Gibbs' entropy on the box size:

$$[S_{\text{Gibbs}}(\epsilon) - S_{\text{ideal}}] = -k \langle \ln(f / f_{\text{ideal}}) \rangle_\epsilon,$$

where S_{ideal} and f_{ideal} refer to the uniform probability density characteristic of an ideal gas.

Other dimensions can be obtained by using formal "measures" which are *powers* of the actual probability, proportional to the number of pairs of points, or triples of points, ... , found in each box. Farmer, Ott, and Yorke suggested that the box-counting and information dimensions have special significance for physically interesting attractors. The box-counting dimension is necessarily unchanged from its equilibrium value in an ergodic system. The information dimension *does* change. It is uniquely important for its connection to dissipation, Liouville's Theorem, the Lyapunov spectrum, and Gibbs' entropy. The "Kaplan-Yorke dimension" D_{KY} is the best estimate for the information dimension in many-dimensional systems with phase spaces too extensive for computing box probabilities. The Kaplan-Yorke dimension is based on a fundamental and natural idea—*a comoving hypervolume having the dimensionality of the attractor can have no long-time tendency to grow or to shrink*. Thus the number of terms at which the linearly-interpolated Lyapunov-exponent sum changes from positive to negative *is* the Kaplan-Yorke estimate for the information dimension $D_I \simeq D_{\text{KY}}$. This correspondence has been proven for the Baker Maps. Because $\langle (\langle p \rangle + \delta p) \ln(\langle p \rangle + \delta p) \rangle$ is *minimal* when the probability fluctuation δp vanishes, D_I is necessarily less than D_{BC} for *multifractal* distributions.

Are the instantaneous Lyapunov exponents and the various fractal dimensions point functions? They *are* for a specified coordinate system—Christoph Dellago and I studied this carefully for both continuous and impulsive motion equations. But Hamiltonian systems can be described by *any* convenient generalized coordinates $\{q\}$ and the corresponding generalized momenta $\{p = \partial \mathcal{L} / \partial \dot{q}\}$. For these systems the sums of all the Lyapunov exponents vanish, and the exponents are pairs $\{(+\lambda, -\lambda)\}$. But, even in this simplest case, the local exponents, like the local fractal dimensions, depend upon the choice of coordinates, and so are properties of the representation, not the system. Nonhamiltonian thermomechanical systems can likewise be described with a variety of physical coordinates. Typically the total time-rate-of-change of the extension in phase space *has* physical significance, $-\dot{\otimes}/\otimes = \dot{S}_{\text{external}}/k$, but the individual exponents need not. Thus the individual local Lyapunov exponents are *not* point functions unless a particular coordinate system is specified. With Nosé-Hoover reservoirs their sum gives the logarithmic rate of phase-volume loss, and the corresponding dissipation, but the local individual contributions to that loss depend upon the chosen coordinate system.

In cases which do not correspond to physical systems, there is no natural alternative set of equations. But new rotated coordinate combinations such as $\{(x + y)/\sqrt{2}, (x - y)/\sqrt{2}\}$ could be introduced, and would lead in general to new values for the local Lyapunov exponents. Lorenz' well-known butterfly-shaped attractor stems from the irreversible motion equations of continuum hydrodynamics. The irreversible nature of those equations is reflected in its localized attractor structure, which is very different to the ergodic repellor-attractor pairs found in reversible dynamical simulations.

7.6   A Simple Ergodic Fractal

Numerical solutions of the time-reversible dynamical equations describing *non*equilibrium thermomechanical problems are *never* time-reversible. The many-to-one dissipative nature of nonequilibrium flows prevents exact computational reversibility. Nevertheless, the necessarily-approximate computer-generated solutions agree on two points: *ergodicity* and *multifractality* are generic characteristics of many dissipative nonequilibrium stationary states, quite independently of the details distinguishing the underlying computer algorithm from its fellows. *Ergodicity* implies that the computational time series approximating the attractor (eventually) covers *every* allowable point while *multifractality* suggests an extremely singular discontinuous structure. How can *both* properties be present simultaneously? This paradox is best understood through an example.

Let us consider sets made up of asymmetric one-dimensional random walks, with the probability of a step to the right arbitrarily chosen to be twice that of a step to the left. Each of the 2^N possible N-step walks can be represented as an N-digit binary number on the unit interval. If probabilities of $\frac{2}{3}$ are associated with the zeros and $\frac{1}{3}$ with the ones in each binary fraction B, the resulting walk probabilities $\{p(B)\}$ are normalized in the usual way:

$$p(B) \equiv (\frac{2}{3})^{N_0}(\frac{1}{3})^{N_1} \; ; \; \sum_B p(B) = \prod_N [(\frac{2}{3}) + (\frac{1}{3})] = [(\frac{2}{3}) + (\frac{1}{3})]^N \equiv 1.$$

Consider now the distribution of probabilities as a function defined on the unit interval. The distribution of walk probabilities is clearly enough a good model for an *ergodic* distribution, with density everywhere and a point-to-point spacing as small as desired, $\Delta B = \epsilon = 2^{-N}$. The probabilities themselves have a wildly-discontinuous fractal character. Figure 7.4 shows the *integrated* cumulative probability $C(B)$:

$$C(B) \equiv \int_0^B p(B')dB'.$$

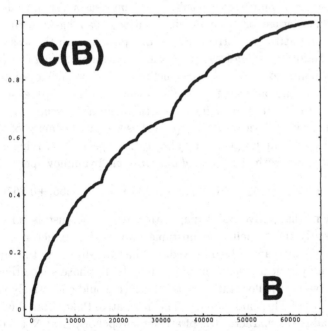

Figure 7.4: *Cumulative* probability $C(B)$ using 65,536 sampling bins. The underlying fractal distribution of random-walk probabilities is "ergodic", with density everywhere, and has an information dimension of 0.918296.

The fractal character of the underlying distribution p(B) can be established by calculating its information dimension D_I from the averaged value of the coarse-grained probabilities,

$$D_I \equiv \langle \ln \mathrm{p}/\ln \epsilon \rangle \equiv [\tfrac{2}{3} \ln \tfrac{2}{3} + \tfrac{1}{3} \ln \tfrac{1}{3}]/\ln \tfrac{1}{2} = 0.918296.$$

Thus the small-ϵ limit of this distribution not only has density everywhere but is also fractal.

7.7   Fractal Attractor-Repellor Pairs

That time-reversible motion equations can lead to irreversible behavior *is* paradoxical. But there are many examples which are above reproach. The time-reversible dissipative Baker map, discussed in Chapter 2, generates a strange attractor with an equilibrium value for the box-counting dimension, which is exactly two, but an information dimension of approximately 1.7337. The same multifractal results from *any* reasonable initial condition, as a time average, or, from any distribution, as a long-time limit. The resulting fractal attractor corresponds to dissipation, with the loss of one-third bit of "information" for every iteration of the map $M(q,p) \to (q',p')$. Should we attempt to recover the past history at any particular point (q,p), by iterating the map backward in time, starting with $M(q,-p)$, exactly the *same* attractor—not its mirror image—is the inevitable result. Nevertheless, the time-reversed set of points, $\{q,-p\}$ *is* the mirror-image of the attractor, an exactly similar geometric object, the *repellor*, satisfying exactly the same map, but with the *reversed and unstable* Lyapunov spectrum,

$$(+\lambda_1 = +0.6365, +\lambda_2 = -0.8676) \longrightarrow \{\lambda_R\} = (-0.6365, +0.8676).$$

The underlying dissipative Baker map *is* time-reversible, just as is its spectrum. Evidently the "repellor" is unstable, just as the attractor is stable. Because the repellor cannot be observed, and has measure zero, the reversed states have no physical significance. In many-body phase space the similarly unobservable repellor states are those which would collectively violate the Second Law of Thermodynamics. The dissipative Baker Map illustrates another paradoxical characteristic of attractor-repellor pairs. Not only are the box-counting dimensions of the attractor and repellor equal. Both objects occupy *exactly the same boxes*. Only the box *weights* are different.

The summed-up attractor weights approach unity while the summed-up repellor weights approach zero as the resolution length ϵ decreases to zero. Evidently the distinction between an "attractor box" and a "repellor box" has no operational significance!

In phase space, continuing collapse, to an attractor, is the path of least resistance for a time-reversible nonequilibrium steady state. Its reverse, sustained expansion, is obviously impossible in a confined space. This important symmetry breaking, caricatured in dissipative maps, is common to all nonequilibrium sets of time-reversible differential equations, such as those used in simulating the Rayleigh-Bénard flows with atoms in Section 4.9.2. Such a flow, when traced backward in time, would converge to an unstable mirror image of the flow forward in time. Because this artificial repellor is actually unstable (in the sense that nearby trajectories are repelled) relative to its parent attractor, it can *only* be generated formally, and artificially, by storing attractor states and playing them backwards. In simulations of systems incorporating heat reservoirs, the relative instability of the repellor revealed by its Lyapunov spectrum corresponds to the impossibility of systems' long-time averages violating the Second Law of Thermodynamics, as detailed in the following Section. The attractor states, generated going *forward* in time, are the only states which can be observed with a finite information supply.

Let us consider a simple time-reversible "attractor", which degenerates to a one-dimensional line in three-dimensional space. The (q, v, ζ) space for a falling thermostated particle illustrated in Section 5.4, is such an example. The equations of motion,

$$\{\dot{q} = v \; ; \; \dot{v} = 1 - \zeta v \; ; \; \dot{\zeta} = v^2 - 1\},$$

are time-reversible, with both the velocity v and the friction coefficient ζ changing sign in the reversed trajectory. The "attractor" for this motion is the line $(q, +1, +1)$. The "repellor" analog is the time-reversed line $(q, -1, -1)$. All possible flows in (q, v, ζ) space are portions of trajectories connecting the two one-dimensional structures.

Time reversibility requires reversing the system boundary velocities too. Thus the repellor corresponding to a steady many-body shear flow, with a *positive* strain rate $dv_x/dy = \dot{\epsilon} > 0$, for instance, differs from the instantaneous time-reversed configuration, which would have a *negative* strain rate incorporated in its boundaries. The attractor and repellor for any specific flow, necessarily occupy the *same* phase space with the *same* boundary conditions including the *same* strain rate $\dot{\epsilon}$. The phase-space repellor for a shear flow with positive strain rate corresponds to the instantaneous time reversal of a *mirror-image* steady shear flow with the *negative* strain rate $dv_x/dy = -\dot{\epsilon} < 0$. The phase-space flow then occurs *from* the repellor *to* the attractor with identical boundary conditions throughout.

7.8   A Global Second Law from Reversible Chaos

Begin with the language and ideas provided by chaos theory—in particular the concepts of fractal distributions and Lyapunov spectra. Then consider a conservative Newtonian system and drive that system away from equilibrium with a specific algorithmic representation of heat reservoirs— reservoirs controlled by Nosé-Hoover thermostat forces. The result is a simple, but compelling, *proof* of a Global *time-averaged* microscopic version of the macroscopic Second Law of Thermodynamics:

> **Long-time-averaged time-reversible nonequilibrium steady- state flows invariably generate external entropy and corre- spond to fine-grained contracting flows *from* a fractal repellor *to* its mirror-image strange attractor.**

The one-way nature of this global Second Law emerges naturally, for computer simulations of simple experiments involving mass flows, shear flows, or heat flows, despite the formal time reversibility of the Newtonian and Nosé-Hoover equations of motion. Two assumptions are required for the proof: (i) *existence* of any long-time-averaged macroscopic quantities of interest; (ii) *independence* of these averaged quantities to the detailed nature of the microscopic initial conditions.

Sources and sinks of energy are necessary to drive or to moderate *any* nonequilibrium flow. In order for heat transfer to occur at least one of these sources and sinks is necessarily a Nosé-Hoover heat reservoir. The two required *non*equilibrium assumptions have corresponding analogs *at* equilibrium: (i) Observable macroscopic quantities exist (so that the sys- tem is stationary and *stable* from the thermodynamic point of view); (ii) Long-time averages converge to the phase-space averages used in Gibbs' statistical mechanics. The two analogous *nonequilibrium* assumptions are most easily motivated by the observation that both laboratory experiments and computer experiments typically (i) give *definite* results, and (ii) these results are *reproducible*, despite differences in the details of the initial conditions and the makeup of the thermostats providing nonequilibrium temperature control.

Although the proof sketched below applies to *any* thermomechanical Newtonian system, the thermal driving of its reservoirs must be provided by the differentiable Nosé-Hoover equations of motion[‡]. With Nosé-Hoover heat reservoirs it is simple to evaluate the phase-space collapse rate from the equations of motion. The calculation follows directly from the local phase-space velocity divergence, $\nabla \cdot v$:

$$-\dot{f}/f \equiv \dot{\otimes}/\otimes \equiv \nabla \cdot v \equiv \sum \lambda_{\mathrm{L}}.$$

It should be emphasized that although the individual *local* Lyapunov exponents $\{\lambda_{\mathrm{L}}\}$ *do* depend upon the chosen coordinate system, *their sum does not*. Only the complete sum is crucial to the argument which follows.

The Nosé-Hoover heat-reservoir forces $\{-\zeta p\}$ describe one or more reservoir regions with specified temperatures maintained by the friction coefficients $\{\zeta\}$. These coefficients—$\{\zeta_{\mathrm{cold}}, \zeta_{\mathrm{hot}}\}$ in the case of a simple two-reservoir heat-transfer simulation—are *independent* variables, so that the phase-space velocity divergence is a simple sum of friction coefficients:

$$\nabla \cdot v \equiv \sum_q (\partial \dot{q}/\partial q) + \sum_p (\partial \dot{p}/\partial p) + \sum_\zeta (\partial \dot{\zeta}/\partial \zeta) = \sum_q 0 - \sum_p \zeta + \sum_\zeta 0.$$

Note that the one nonzero sum $\sum_p \zeta$ includes only those degrees of freedom belonging to thermostated regions. Independence of averages to the initial conditions implies that exactly the same expression necessarily holds as a *time average*, giving the *averaged* comoving velocity divergence $\langle \nabla \cdot v \rangle$ in terms of the long-time average of the summed-up friction coefficients. In the "extended" (q, p, ζ) phase space Liouville's *compressible* Theorem applies. It relates the changing probability density, *following the flow*, to the velocity divergence and so to the local Lyapunov exponents $\{\lambda_{\mathrm{L}}\}$ and the changing extension in phase $\otimes$:

$$d \ln f/dt \equiv -\dot{\otimes}/\otimes \equiv -\nabla \cdot v = +\sum_p \zeta = -\sum_\lambda \lambda_{\mathrm{L}} \longrightarrow$$

$$\langle d \ln f/dt \rangle \equiv -\langle \dot{\otimes}/\otimes \rangle \equiv -\langle \nabla \cdot v \rangle = +\langle \sum_p \zeta \rangle = -\langle \sum_\lambda \lambda_{\mathrm{L}} \rangle.$$

[‡]Posch and Dellago have developed special techniques for use with impulsive forces.

It is crucial to see that the long-time averages of each of the five terms in these equations *must* be positive, corresponding to *collapse* (of comoving phase-space hypervolume), rather than negative, which would correspond to divergence. To see this, consider the time development of a small compact hypervolume $\otimes$ in the extended phase space. For definiteness imagine the hypervolume to be a hypersphere initially containing a constant normalized density function,

$$f(q, p, \zeta, t = 0) \equiv f_0 = 1/\otimes,$$

for all those points "inside" the hypervolume $\otimes$. This fine-grained probability density has an integral of unity and vanishes outside $\otimes$. It corresponds physically to a *nonequilibrium ensemble* of similarly-prepared systems.

Now apply Liouville's *compressible* Theorem. According to that theorem, the local expansion or contraction of the hypervolume and the local increase or decrease of the density $f(q, p, \zeta, t)$ responds to the local summed-up friction coefficients $\sum_p \zeta$. The small hypervolume, initially a hypersphere, will rapidly be transformed to a hyperellipsoid. The corresponding initial growth and decay rates are described by the local Lyapunov exponents. At somewhat longer times (of order $q/\dot{q} \simeq p/\dot{p}$) the small ellipsoid bends and stretches. The local growth rate and density become nonuniform throughout the moving hypervolume. At *much* longer times—of order $(\ln L)/\lambda_1$—the fibrillating hypervolume element traverses the space. At *very* long times we have already assumed that the distribution becomes *uniform*, in a sufficiently coarse-grained sense, making it possible to calculate convergent steady averages *independent* of the initial location of the hyperspherical extension in phase $\otimes$:

$$\langle \dot{\otimes}/\otimes \rangle_{\text{local}} \equiv \langle \dot{\otimes}/\otimes \rangle_{\text{global}}.$$

Throughout its time development the comoving "fine-grained" density responds to the summed-up local friction coefficients as is described by the summed-up local Lyapunov exponents. Consider now the time-*averaged* density change following a particular solution of the equations of motion for a time t. According to Liouville's Theorem, applied to a nonequilibrium Newtonian system interacting with one or more Nosé-Hoover heat reservoirs:

$$d \ln f/dt \equiv \sum_p \zeta \longrightarrow f(t) = f_0 e^{\int_0^t \sum \zeta \, dt'} \equiv f_0 e^{t \langle \sum \zeta \rangle}.$$

Because it is assumed that the boundary conditions are fixed, so that a *steady* state results[§], there are only three possibilities for the long-time limit of such a comoving fine-grained density:

$$\langle \ln f_{\text{steady}}(t \longrightarrow \infty) \rangle \longrightarrow \{-\infty, \ln f_0, +\infty\}.$$

These three cases correspond to three possibilities for the time-averaged friction-coefficient sum:

$$\{\langle \sum \zeta \rangle < 0 \,,\, \langle \sum \zeta \rangle = 0 \,,\, \langle \sum \zeta \rangle > 0\}.$$

The *first* possibility, that $\langle \sum \zeta \rangle$ is negative, implies, through Liouville's Theorem, the *divergence* of the comoving volume element (which has to be independent of the hypersphere's initial location), which is inconsistent with a convergent stable solution. Thus the first "possibility" must be ruled out.

The *second* possibility, that $\sum \zeta$ has a stable average value of 0, *must* correspond to equilibrium. This is obvious for the simplest case, a Newtonian system coupled to exactly two heat reservoirs, one at T_{cold} and one at T_{hot}. Then the zero-sum condition,

$$\langle \sum_p \zeta \rangle = \langle (\#\zeta)_{\text{cold}} \rangle + \langle (\#\zeta)_{\text{hot}} \rangle = 0,$$

is inconsistent with the long-time-averaged *energy-balance* condition:

$$\langle (\#\zeta)_{\text{cold}} T_{\text{cold}} \rangle + \langle (\#\zeta)_{\text{hot}} T_{\text{hot}} \rangle = 0,$$

unless the two temperatures are equal, in which case the averaged heat transfers necessarily vanish. We conclude that a simple system *with* heat transfer between two heat reservoirs can have neither a negative nor a vanishing friction-coefficient sum. There seems to be no direct argument to show that the zero-sum condition must correspond to equilibrium when more than two reservoirs are present.

Only the *third* and final possibility, $\langle \sum \zeta \rangle > 0$, remains. This possibility is thereby necessarily established, away from equilibrium, and implies that *the Gibbs entropy approaches* $-\infty$ *in a nonequilibrium steady state*:

$$\langle (\dot{S}_{\text{Gibbs}}/k) \rangle \equiv -\langle d \ln f/dt \rangle = \langle \dot{\otimes}/\otimes \rangle = -\langle \sum \zeta \rangle < 0 \Longrightarrow$$

$$(S_{\text{Gibbs}}/k) \to -\infty !$$

[§]A similar result also holds if the boundary conditions vary periodically with time.

This divergence of the Gibbs entropy corresponds to the vanishing fraction of the phase space effectively occcupied by the attractor's "core"[¶],

$$\langle\otimes\rangle_{\text{steady}} \simeq \langle\otimes\rangle_{\text{equilibrium}}^{D_I/D_E},$$

where D_E is the "equilibrium" or "embedding" dimensionality, that of the full phase space.

For two heat reservoirs the proof just sketched establishes that $\langle\zeta_{\text{hot}}\rangle$ is *necessarily* negative, and $\langle\zeta_{\text{cold}}\rangle$ is *necessarily* positive, corresponding to the usual flow of heat, from hot to cold. With more than two heat reservoirs, the various directions of the individual flows cannot generally be determined in advance, but it must *necessarily* still be true that the full friction-coefficient sum, over all thermostated degrees of freedom, can only have a positive long-time average, corresponding to the attractive collapse of phase-space density onto an attractor. Let us reïterate the conclusion, the global Second Law of Thermodynamics:

Long-time-averaged time-reversible nonequilibrium steady-state flows invariably generate external entropy and correspond to fine-grained contracting flows *from* a fractal repellor *to* its mirror-image strange attractor.

Although the two divergences, of the "fine-grained" phase-space density and the corresponding Gibbs entropy, seemed odd in 1986, when the first simulations studying this effect were carried out, a variety of succeeding low-dimensional simulations confirmed a relatively-simple geometric interpretation. In every case, the limiting steady-state probability density $f(q, p, \zeta)$ collapses and diverges on a fractal "strange attractor",

$$\langle f\rangle \to +\infty \; ; \; \langle\otimes\rangle \to 0 \; ; \; S_{\text{Gibbs}} = -k\langle\ln f_N\rangle \to -\infty.$$

The Kaplan-Yorke estimate of the attractor's "information dimension" is also *necessarily* strictly less than that of the original equilibrium extended phase space.

The information dimension D_{I} reflects the way in which the coarse-grained entropy,

$$S_{\text{CG}}(\epsilon) = -k\langle\ln p_{\text{CG}}\rangle_\epsilon,$$

[¶]Farmer, Ott, and Yorke so term the physically-significant high-probability region.

depends upon the box size ϵ and the corresponding coarse-grained probabilities $\{p_{CG} \propto \epsilon^{D_I}\}$ (for small ϵ), as we saw in Section 7.5:

$$D_I \equiv \langle \ln p_{CG} \rangle_\epsilon / \ln(\epsilon) \equiv \sum p_{CG} \ln p_{CG} / \ln(\epsilon).$$

The fractal explanation of the probability density's divergence has additional physical significance. *It indicates the extreme rarity of nonequilibrium states.* These core states have *zero measure* relative to the continuously-distributed equilibrium states. This means that any fixed fraction of the total nonequilibrium measure (half, nine tenths, ninety nine percent, ...) can be found within a *vanishing* fraction of the total number of boxes in the limit that the box size approaches zero. The averaged rates of divergence of $\ln f$ and S_{Gibbs} correspond also to a (long-time-averaged) *negative* Lyapunov sum, so that any comoving hypervolume element eventually collapses, with a volume of order $e^{-\Delta S/k}$, where ΔS is the time-integrated net entropy increase of the external Nosé-Hoover heat reservoir(s). Suppose now, as is usual, that the equations of motion are time-reversible. Then, the time-reversed trajectory, with the momenta and friction coefficients changed in sign, has a *positive* time-averaged Lyapunov sum, corresponding to instability and a diverging hypervolume element.

Such hypothetical time-reversed trajectories cannot be observed, even though they are formally legitimate "solutions" of the equations of motion. This symmetry breaking, with forward attractor trajectories collapsing in a stable way and reversed repellor trajectories unstable and unobservable, *is* the mechanical equivalent of the Second Law of Thermodynamics. Before considering detailed examples, let us briefly consider the relation between the fine-grained entropy change from Liouville's compressible theorem and the macroscopic coarse-grained entropy production discussed in Sections 4.5 and 5.7.

7.9 Coarse-Grained and Fine-Grained Entropy

For a nearly homogeneous system described by Newtonian viscosity and Fourier's heat conduction, a phenomenological system-entropy production,

$$\dot{S}_{\text{prod}} = [(\eta/T)\dot{\epsilon}^2 + \kappa(\nabla \ln T)^2]V,$$

is often arbitrarily introduced in order to avoid any net entropy change *within* a stationary nonequilibrium state. Such an internal macroscopic entropy production should be chosen to offset the *decrease* with time of the *fine-grained* Gibbs' entropy discussed in the last Section:

$$\langle \dot{S}_{\text{Gibbs}} \rangle \equiv -kd\langle \ln f \rangle / dt.$$

The continual evolution of the fine-grained entropy in a "steady" state (!) describes the penetration of information and phase-space structure to smaller and smaller scales. Because these scales have no physical significance below the limits of observation, and further have no computational significance below the level of precision carried, it is tempting to define a "coarse-grained" entropy based on an averaged probability density function. Because we assume the accurate convergence of macroscopic quantities at long times, the corresponding coarse-grained entropy $S_{\text{CG}}(\epsilon)$ must eventually converge too, for any fixed box size ϵ. This coarse-grained entropy has the substantial drawback that its box-dependent definition is quite arbitrary, and is even unbounded from below, away from equilibrium.

Such a coarse-grained entropy has sometimes been used to fill the perceived need for an entropy away from equilibrium. In his boxed notes at the Yale University library Gibbs repeatedly muses over such a nonequilibrium entropy. He even mentions two specific examples, steady flows with density or temperature gradients. Evidently Gibbs could find no convincing answer to this question. If there *were* a well-defined nonequilibrium entropy we could perhaps avoid the fractal divergence of the Gibbs entropy. Gibbs' fine-grained entropy inexorably *decreases* toward minus infinity in nonequilibrium steady states. To save the notion of a finite fixed steady-state entropy, it is necessary to somehow compensate for the actual decrease. In continuum mechanics, entropy *production* is traditionally introduced, arbitrarily, for this reason. It is evident, with the coarse-grained entropy fixed and constant, while Gibbs' fine-grained entropy decreases, that the phenomenological entropy production can be defined so as to exactly cancel the loss of Gibbs' entropy:

$$0 \equiv \langle \dot{S}_{\text{CG}} \rangle \equiv \langle \dot{S}_{\text{Gibbs}} + \dot{S}_{\text{prod}} \rangle \longrightarrow \langle \dot{S}_{\text{prod}} \rangle = -\langle \dot{S}_{\text{Gibbs}} \rangle.$$

In a series of very interesting papers, Tél, Vollmer, and Breymann have extended this idea to open systems, in which an additional entropy change, due to mass flows, is included. See the recent literature, as well as their paper in Europhysics Letters **35**, 659-664 (1996).

By connecting the fine-grained change of Gibbs' entropy with the phenomenological entropy production of irreversible thermodynamics, we see that the microscopic approach emphasizes the loss of extension in phase due to nonequilibrium constraints while the macroscopic approach ascribes the external entropy change of heat reservoirs to an internal entropy production. To clarify these different points of view I discuss four example problems here. First, I consider a chaotic double pendulum, and point out that the local Lyapunov spectrum for this system depends upon the coordinate system used to describe it, while the global spectrum does not. This equilibrium example is followed by three nonequilibrium problems illustrating the generic features of the thermomechanical Second Law of Thermodynamics stated in the previous Section.

7.10   Example Problems

The problems considered here illustrate Lyapunov spectra for equilibrium and nonequilibrium systems. At equilibrium, the planar Hooke's-Law double pendulum is confined to a seven-dimensional energy surface in its eight-dimensional phase space (four coordinates and four momenta define the space). It is instructive, as was stressed in Chapter 1, to see that the instantaneous local contributions to the Lyapunov spectrum differ for two "natural" choices of coordinates, Cartesian and polar. In the second problem I illustrate the connection between fractal distributions and the divergence of Gibbs' entropy for the Galton Board problem. Here, the two-dimensional phase-space Poincaré section cataloging the possible collisions is easy to visualize. It is relatively easy to show ergodicity for this system numerically. Ergodicity was later *proved* by Chernov, Eyink, Sinai, and Lebowitz. Enhancing the resolution of the Poincaré section displays the inexorable approach of the Gibbs' entropy to $-\infty$.

Next we consider a heat-conducting harmonic oscillator. It has a four-dimensional strange attractor, sufficiently complex to illustrate symmetry-breaking in the Lyapunov spectrum and the lack of a simple hyperbolic structure. For small deviations from equilibrium, the *symmetric* equilibrium structure of the spectrum $\{+\lambda, 0, 0, -\lambda\}$ is replaced by an *asymmetric* spectrum, with one positive Lyapunov exponent and *two* negative exponents. Finally, we consider the "color conductivity" problem, illustrating the shift of Lyapunov exponents to more negative values in a many-body system. In the time-reversible *non*equilibrium problems considered here this shift of Lyapunov exponents toward more negative values is directly related to phase-space contraction. In the Galton Board and Color Conductivity problems the phase-space contraction rate is *identical* to the external rate of thermodynamic dissipation. Klages, Rateitschak, and Nicolis have emphasized that this identity need not hold in more complicated cases. In *every* case, the resulting fractal distributions also explain, quantitatively, the general rarity of nonequilibrium phase-space states relative to their much more numerous equilibrium relatives.

7.10.1 *Chaotic Double Pendulum*

Consider the chaotic Hooke's-Law *double* pendulum problem, with both masses unity, but with variable pendulum lengths, in a unit gravitational field. We arbitrarily choose the force constants in the two Hooke's-Law springs equal to four. The equations of motion are simply related to those of the single pendulum considered in Section 1.8.3:

$$\dot{x}_1 = p_{x1} \; ; \; \dot{y}_1 = p_{y1} \; ; \; \dot{x}_2 = p_{x2} \; ; \; \dot{y}_2 = p_{y2} \; ;$$

$$\dot{p}_{x1} = 4(x_1)[(1/r_1) - 1] + 4(x_1 - x_2)[(1/r_{12}) - 1] \; ;$$

$$\dot{p}_{y1} = 4(y_1)[(1/r_1) - 1] + 4(y_1 - y_2)[(1/r_{12}) - 1] - 1 \; ;$$

$$\dot{p}_{x2} = 4(x_2 - x_1)[(1/r_{12}) - 1] \; ; \; \dot{p}_{y2} = 4(y_2 - y_1)[(1/r_{12}) - 1] - 1.$$

Whether or not a particular solution is chaotic depends solely on the initial conditions. Small displacements from the least-energy configuration, $y_1 = -3/2, y_2 = -11/4$, give four normal vibrational modes and no chaos. High energies eliminate the influence of the gravitational field. Again no chaos results.

The initial horizontal unstressed configuration shown in the Figure, with no initial kinetic energy, *is* chaotic. The distributions of the *local* values of the four largest Lyapunov exponents, $\{-5 < \lambda < +5\}$, calculated in both the Cartesian phase space, $\{x, y, p_x, p_y\}$, and the polar phase space, $\{r, \theta, p_r, p_\theta\}$, are shown in Figure 7.5. In both cases the averaged exponents are $\{+0.17, +0.08, +0.04, 0.00\}$. The constant-energy condition corresponds to the vanishing exponent. A *second* paired zero exponent corresponds to motion in the direction of the trajectory motion. Note that the corresponding *instantaneous* exponents do not vanish. The local exponents *do* satisfy the exact instantaneous pairing rules, $\{\lambda_i + \lambda_{9-i} \equiv 0\}$.

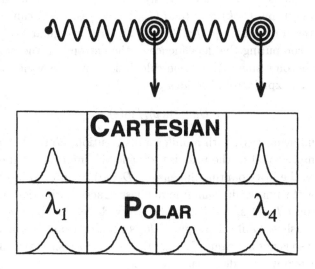

Figure 7.5: Distributions of double Hookean pendula spectra in Cartesian and Polar phase spaces. Gravity accelerates both masses downward. Distributions $\{p(\lambda \geq 0)\}$ for the four most-positive exponents mirror the $\{p(\lambda \leq 0)\}$ for the four most-negative exponents not shown here. For each exponent the range of values shown is $-5 < \lambda < +5$.

7.10.2 *Coarse-Grained Galton Board Entropy*

The singular multifractal nature of strange attractors can be illustrated using the isokinetic Galton Board example worked out in Section 5.9.1. In the equilibrium case, with zero accelerating field, the probability density is

Gibbs' uniform microcanonical distribution:

$$f_{eq}(0 < \alpha < \pi, -1 < \sin \beta < +1) = 1/(2\pi),$$

corresponding to the equal probability of the direction of motion and the location of scattering collisions. A nonvanishing accelerating field changes the uniform density in an intrinsically singular way—the mean value of the excess entropy, $(S/k) = -\langle \ln(f/f_{eq}) \rangle$ diverges. This can be seen by spanning the unit square $[(\alpha/\pi), (\sin\beta)/2)]$ with a regular grid of square boxes, and computing the dependence of the entropy on the box size. As the boxes become small, this same calculation gives also the information dimension, as explained in Sections 2.11 and 7.5:

$$D_I = \langle \ln p \rangle_\epsilon / \ln(\epsilon).$$

In the equilibrium case, with a uniform distribution, *all* the box probabilities are simply equal to the box size, ϵ^2, and the information dimension is the same as the box-counting dimension, $D_I(eq) = \langle \ln(\epsilon^2) \rangle / \ln(\epsilon) \equiv 2$.

Because the nonequilibrium fractal phase-space distribution contains information on *all* scales, and is singular everywhere, the effective extension in phase vanishes. Gibbs' entropy S_G for such a nonequilibrium system diverges, and so is not a meaningful quantity. Figure 7.6 shows the divergence of Gibbs' entropy with decreasing box-size ϵ.

$$D_I = \langle (\ln f\epsilon^2) \rangle_{CG} / \ln(\epsilon) = \langle \ln p \rangle_{CG} / \ln(\epsilon) = -[S_G/k \ln(\epsilon)] + 2.$$

Here "coarse-grained" averages are indicated by a subscript $_{CG}$.

Figure 7.6: Dependence of Gibbs' entropy on the box size ϵ. The linear relation, $(S_G/k) = +0.18 \ln \epsilon < 0$, indicates an information dimension of 1.82. Details are in my paper in the 1998 Journal of Chemical Physics.

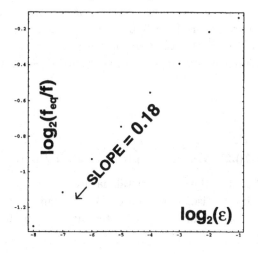

The information dimension just calculated agrees perfectly with Dellago's independent calculation of the Kaplan-Yorke dimension from the Lyapunov spectrum. See his 1995 PhD thesis (University of Vienna) for details. Gary Morriss has also investigated this same system, and all three sets of results are in good agreement. It is instructive that in Morriss' work on the Galton Board, he introduced different phase-space coordinates, a different field direction, different density and distance scales, and a different name for the problem, all in the pursuit of an imaginary novelty!

7.10.3 *Heat-Conducting Harmonic Oscillator*

A thermostated harmonic oscillator requires *two* time-reversible control variables, ζ and ξ, to reproduce Gibbs' canonical distribution. For an oscillator with unit mass, force constant, and temperature, Gibbs' canonical distribution is:

$$f(q,p)_{\text{eq}} = (2\pi)^{-1} e^{-(q^2+p^2)/2}.$$

If the thermostated equations of motion are these:

$$\{\dot{q} = p \ ; \ \dot{p} = -q - \zeta p \ ; \ \dot{\zeta} = p^2 - 1 - \xi\zeta \ ; \ \dot{\xi} = \zeta^2 - 1\},$$

the extended phase-space distribution, in the four-dimensional (q, p, ζ, ξ) space,

$$f(q, p, \zeta, \xi) = (2\pi)^{-2} e^{-(q^2+p^2+\zeta^2+\xi^2)/2},$$

satisfies the stationarity condition $(\partial f/\partial t) \equiv 0$. A numerical solution of five neighboring trajectories (twenty differential equations in all) gives the expected symmetric Lyapunov spectrum $\{+0.066, 0.000, 0.000, -0.066\}$.

This oscillator problem becomes a *non*equilibrium problem if the temperature T is nonuniform. This allows overall dissipation through hot-to-cold heat transfer. A numerical solution of five neighboring trajectories of the *non*-equilibrium motion equations,

$$\{\dot{q} = p \ ; \ \dot{p} = -q - \zeta p \ ; \ \dot{\zeta} = p^2 - T(q) - \xi\zeta \ ; \ \dot{\xi} = \zeta^2 - T(q)\};$$

$$T(q) \equiv 1 + 0.25 \tanh(q),$$

has a spectrum with *two* negative exponents: $\{+0.053, 0.0, -0.035, -0.086\}$. Despite the time-reversible equations of motion the possibility of dissipation is, as usual, irresistable, making overall contraction inevitable. The distribution accordingly converges to a strange attractor, with a Kaplan-Yorke dimension $3 + (0.018/0.086) = 3.21$.

The four offset vectors $\{\delta_1, \delta_2, \delta_3, \delta_4\}$ separating the reference trajectory from its four satellite trajectories reveal a *second* symmetry breaking. Despite the exact time-reversibility of the equations for the Lyapunov spectrum, the vectors reflect their *past* history *independent of their futures*. Playing the reference trajectory backward (changing the time-ordering of the points, as well as the signs of p, ζ and ξ) reveals that the sum of the local Lyapunov exponents changes sign but that the individual Lyapunov exponents behave in a more complicated way. The sum is $-\zeta - \xi$. If the satellite trajectories were themselves played backward, then the instantaneous Lyapunov spectrum *would* simply change sign. But if the vectors are not so constrained, a *second* symmetry breaking appears. *Unless* the initial point is chosen common to *both* the attractor and the repellor—$(p, \zeta, \xi) = (0, 0, 0)$ is such a point—the instantaneous forward and backward spectra *at* the initial point will generally be found to differ.

7.10.4   *Color Conductivity*

The Galton Board problem of Section 7.10.2 follows the motion of a *single* particle through a lattice of fixed scatterers. This one-body problem is isomorphic to a *two*-body problem, with periodic boundaries, in which the two particles are accelerated, in opposite directions, by a fixed external field. In this form it is natural to generalize the two-particle version of the Galton Board problem to N particles, with $N/2$ of each type, accelerated in opposite directions. This "Color Conductivity" problem is the many-body analog of the Galton Board. In this many-particle case the equations of motion have the form:

$$\{\dot{x} = (p_x/m) \; ; \; \dot{y} = (p_y/m) \; ; \; \dot{p}_x = F_x \pm E - \zeta p_x \; ; \; \dot{p}_y = F_y - \zeta p_y\},$$

where we arbitrarily choose the field E in the x direction. It is natural, and usual, to choose the friction coefficient ζ to thermostat the total kinetic energy, making K a constant of the motion:

$$(d/dt)[\sum_{\text{white}} (p^2/2m) + \sum_{\text{black}} (p^2/2m)] \equiv 0.$$

Generally, the friction coefficient ζ can be chosen to stabilize the kinetic or internal energy of the system, or can simply be chosen equal to a convenient constant. This latter choice corresponds to a pervasive cold stochastic thermostat, with $T_{\text{bath}} = 0$. In either case the current leads to a well-defined conductivity, the "color conductivity", given by the ratio of the mean velocity in the field direction divided by the field strength.

For simplicity, we minimize numerical errors by using a short-ranged purely-repulsive potential, with a characteristic energy ϵ and three continuous derivatives at the cutoff distance, $r = \sigma$:

$$\phi(r < \sigma) \equiv 100\epsilon[1 - (r/\sigma)^2]^4.$$

The length σ is an effective "collision diameter".

At the lower of two field values, $E = 0.25\epsilon/\sigma$, the large-system steady distribution of the two particle "colors" turns out to be spatially homogeneous. The smooth dependence of the results on system size indicates that, like shear viscosity, the color conductivity in two dimensions need not exhibit any hydrodynamic instability. At the higher field value, $0.50\epsilon/\sigma$, the situation is different, with the two colors separating and the "fluid" *freezing*. It is clear from these data that a nonequilibrium phase transition analogous to the coexisting morphologies seen in the Rayleigh-Bénard simulations of Section 6.7, separates the two field strengths. The Lyapunov exponents for the higher-field systems are significantly smaller, indicating the collective motion, with a corresponding reduction in mixing activity in the phase space.

An unexpected and significant finding emerged from an analysis of these Lyapunov instability studies. We found that those particles which make the largest contribution to the maximum exponent, λ_1, tend to be localized in space. Figures 7.7 and 7.8, for $N = 25,600$, illustrate typical cases, following the 1998 paper by Hoover, Boercker, and Posch. Figure 7.7 shows the homogeneous mixed arrangement of the two "colors" which prevails at the lower field strength. Figure 7.8 shows, as larger dots, those particles which make an above-average contribution to the local Lyapunov exponent λ_1, first in coordinate space, and then in momentum space. The correlated clumps of "important particles" which result are nearly the same for the two representations. We found no particular properties of the important particles, such as temperature, energy, or stress, which were correlated with the instability. The very smallest nonzero Lyapunov exponents exhibit a much more regular behavior, with a spatial periodicity, like sound waves. This problem is currently under active investigation by Posch's group in Vienna.

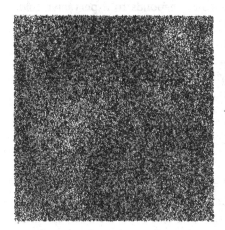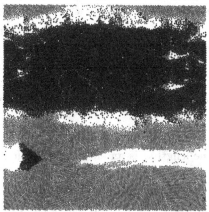

Figure 7.7: The distributions of the two particle "colors" are shown for the two different field strengths discussed in the text. At the higher field strength, shown at the right, the two-component fluid has frozen into crystallites—the distribution is inhomogeneous, with similarly-colored particles clumped together.

Numerical studies of the Lyapunov exponents also reveal, as suggested by Green and Kubo's work, that the distribution has a (fractal) "information dimension", D_I, which lies below the equilibrium embedding dimension, D_E, by a "dimensionality loss" ΔD *quadratic* in the field strength. Because the sum total of the Lyapunov exponents *must* necessarily be negative, and of order $(\epsilon/\sigma)^2$ for small fields, the dimensionality loss can be estimated:

$$\Delta D \equiv D_E - D_I = \langle \dot{S}_{\text{external}}/(k\lambda_1) \rangle.$$

For the fluid example problem illustrated in Figure 7.7 the loss is about 170, where the total phase-space dimensionality is $4 \times 25,600 = 102,400$.

The external dissipation rate due to the thermostat forces is equal to the energy extracted from the system divided by the thermostat temperature.

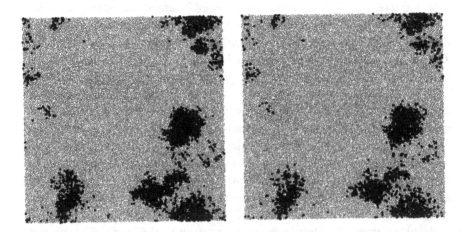

Figure 7.8: In the homogeneous lower-field case illustrated at the left side of Figure 7.7, those particles making above-average contributions to the largest *instantaneous* Lyapunov exponent are emphasized here by using larger dots. The coordinate-space contributors are emphasized at the left; the momentum-space contributors are emphasized at the right.

7.11   Summary

The chaos—localized exponential instability—underlying the fractal attractors characteristic of time-reversible nonequilibrium steady states provides a clear interpretation of the dissipation underlying the Second Law of Thermodynamics. As extraneous information is destroyed by heat transfer, $T\dot{S}_{\text{external}}$, the logarithmic extension in phase volume vanishes. Gibbs' entropy approaches $-\infty$ as the phase-space probability density diverges. The stability of the resulting singular set of states—the set is a strange and multifractal attractor—is to be contrasted with the *instability* of its time-reversed image, the repellor. The Lyapunov Spectrum provides a detailed description of the time-reversible phase-space flow—*from* the unobservable repellor *to* the inevitable attractor—as well as Kaplan and Yorke's connection to the fractal geometry of the attractor and to the time-averaged rate of external entropy increase. The thermomechanical form of the Second Law of Thermodynamics links the external and Gibbs entropies to the collapse of the phase-space distribution function, as given by the long-time-averaged friction coefficients and the Lyapunov spectrum:

$$\langle \dot{S}_{\text{external}}/k \rangle = -\langle \dot{S}_{\text{Gibbs}}/k \rangle = \langle d\ln f_N/dt \rangle = -\langle \dot{\otimes}/\otimes \rangle = \langle \sum \zeta \rangle = -\sum \lambda > 0.$$

$$f_{\text{CG}} \propto \epsilon^{D_I - D_E} \simeq \epsilon^{-\dot{S}_{\text{ext}}/\lambda k} \longleftrightarrow \otimes_{\text{CG}} \propto \epsilon^{D_E - D_I} \simeq \epsilon^{+\dot{S}_{\text{ext}}/\lambda k}.$$

Long-time-averaged time-reversible nonequilibrium steady-state flows invariably generate external entropy and correspond to fine-grained contracting flows *from* a fractal repellor *to* its mirror-image strange attractor.

Chapter 8

Resolving the Reversibility Paradox

It's a Naïve Domestic Burgundy
without any Breeding, But I think
You'll be Amused by its presumption.

James Thurber

8.1 Introduction

Our exploration of time reversibility from the perspective of computer simulation and chaos has provided us with insights into the breaking of time symmetry which were not available to Boltzmann or Gibbs or Maxwell or Poincaré or to their successors: Green, Kubo, and Onsager. Insights gleaned from careful computer simulations have also provided grist for the mathematicians' mills. The main results are already quite clear, despite the lack of formal proofs. Simulations have clarified the formation and significance of time-reversible ergodic multifractal phase-space structures. Those structures, which arise from dissipative chaos, provide a natural link between an underlying time-reversible microscopic dynamics and the one-way irreversibility of the macroscopic Second Law of Thermodynamics. The various approaches to understanding are necessarily consistent with this view, but they are often expressed in quite different languages, sometimes contrived. By searching through libraries, attending topical conferences, and exchanging ideas with my colleagues, I have sought to assimilate these different routes to understanding with my own. I am very grateful to those many who have helped.

Understanding evolving views, but always with a perspective grounded in the past, is the continuous pursuit of a moving target. Only a partial understanding can result. My own point of view reflects in part those of my predecessors and contemporaries. Though consensus is necessarily incomplete in research, my goals here are first, to emphasize common features, and second, to point out useful directions in which new insights leading to a more complete understanding might be found. Wherever disagreement or uncertainty remains, I reïterate that the most powerful and useful means to understanding is that based on the study of relatively simple computational models. This approach, gaining understanding through the analysis of "computer experiments", is far from being exhausted.

8.2 Irreversibility from Boltzmann's Kinetic Theory

Even today, an understanding of Boltzmann's views is complicated by the sheer bulk of the work he generated during the lifelong evolution of his ideas. His language and notation are also unfamiliar—he often uses $\{q\}$ for momenta and $\{p\}$ for coordinates, for example. Steve Brush, Carlo Cercignani, Ezechiel Cohen, Martin Curd, and Giovanni Gallavotti have all provided useful guideposts to the chronology and development of his work. In the early 1960s, while he was still at Livermore, Steve made me a very welcome gift—his translation of Boltzmann's gas-theory lectures. Boltzmann was always interested in clarifying and exploring precise detailed consequences of well-defined mechanical models and was, for this reason, unable to do much with liquids and solids. Even air, with its missing vibrational heat capacity, was already too much for the simple classical mechanical model. Kinetic theory was nonetheless a superb model for the low-density transport coefficients. It was a welcome supplement to continuum mechanics, which is silent on constitutive properties.

Boltzmann's irreversible equation for the time evolution of the single-particle distribution function $f_1(r, v, t)^*$, due to its statistical assumption for the two-body collision frequencies, provides a direct, though approximate, link between two-body atomistic mechanics and the accepted many-body flow equations of continuum mechanics. Despite its approximate na-

---

$^*\dot{f}_1 = (\partial f/\partial t)_{\text{coll}}$; f_2 is given by a product of one-body functions in the collision term.

ture, the Boltzmann equation really does *apply* to (classical models of) dilute gases. The exponential amplification of perturbations justifies his simplification for the relative probability of finding two particles in a small volume element:

$$f_2(r_1, v_1, r_2, v_2, t)\delta(r_1 - r_2) \longrightarrow f_1(r_1, v_1, t)f_1(r_2, v_2, t)\delta(r_1 - r_2).$$

For dilute gases, solutions of the Boltzmann equation provide useful recipes for computing the diffusion coefficient, the shear viscosity, and the heat conductivity. The Boltzmann equation provides an understanding too of the chaotic collisional mechanism underlying the Maxwell-Boltzmann distribution. Best of all, a rigorous consequence of the (approximate) Boltzmann equation for $f_1(r, v, t)$, is an expression for the time-dependence of the corresponding single-particle "Boltzmann entropy":

$$S_{\mathrm{B}}(t) \equiv -Nk\langle \ln f_1 \rangle = -k \int \int dr dv f_1 \ln f_1.$$

Boltzmann's entropy rigorously obeys the usual macroscopic thermodynamic form of the Second Law, *without* the thermal fluctuations that necessarily accompany microscopic dynamical simulations. According to the Boltzmann equation his entropy S_{B} *invariably* increases with time, in isolated systems. This strong result, called the H Theorem, is a direct consequence of the plausible, but approximate, nature of Boltzmann's derivation. Boltzmann replaced Liouville's exact time-*reversible* evolution equation, $\dot{f}_N \equiv 0$, with his own *irreversible* one, $\dot{f}_1 = (\partial f/\partial t)_{\mathrm{coll}}$.

Let us check on the time-reversibility of Boltzmann's Equation by imagining the consequences of reversing all the velocities in a many-body system at some particular time. To explore the consequences, take a distribution $f(r, v, t)$ and change the signs of all the particle velocities, $\{+v\} \rightarrow \{-v\}$. Then reverse the time ordering of points along all the particle trajectories at the chosen reversal time, $\Delta t \rightarrow -\Delta t$. Evidently these two operations *change* the sign of the lefthand side of the Boltzmann equation, $+\dot{f}_1(+v) = -\dot{f}_1(-v)$. The collision term on the righthand side is *unchanged* by the time-reversal operation. Uncorrelated collisions *always* bring a velocity distribution closer to local equilibrium. To see this in detail, notice that each time-reversed two-body collision contributing to the righthand side can be replaced by its "reversed inverse" collision, with both the relative separation *and* the velocity changing sign $(+r_{ij}, +v_{ij}) \rightarrow (-r_{ij}, -v_{ij})$. (The macroscopic location of *any* colliding particle pair is unchanged during their collision, but the microscopic geometry of each colliding pair is inverted.) For every "forward-collision" velocity pair $\{(+v_i, +v_j) \to (+v_i', +v_j')\}$, inverting both the relative separation vector and the velocities gives the time-reversed and spatially-inverted collision $\{(-v_i, -v_j) \to (-v_i', -v_j')\}$. See Figure 8.1 for an example collision.

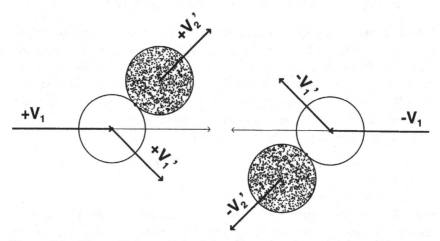

Figure 8.1: The collision on the left: $\{(+v_1, +v_2) \to (+v_1', +v_2')\}$ corresponds exactly [just rotate the drawing 180 degrees] to the time-reversed and spatially-inverted collision on the right: $\{(-v_1, -v_2) \to (-v_1', -v_2')\}$.

Thus the collision term on the righthand side, evaluated for the reversed velocities, is unchanged: $(\partial f[+v]/\partial t)_{\text{coll}} = (\partial f[-v]/\partial t)_{\text{coll}}$. With the lefthand side changing sign, and the righthand side not, we see that the Boltzmann equation *is* intrinsically irreversible. This lack of time reversibility only became clear as objections to the irreversibility were raised. Attempts to rationalize this irreversibility, in the face of Zermélo and Loschmidt's objections to it, engaged Boltzmann's efforts for decades.

Boltzmann finally attempted to avoid Zermélo's recurrence objection and Loschmidt's reversibility objection by imagining a built-in cosmological irreversibility based on an unlikely low-entropy fluctuation, or initial condition, with a resemblance to our current "Big-Bang" "understanding" of how things began. A cosmological resolution of the reversibility paradox was appealing to Boltzmann, and later to the Ehrenfests, and still has an avid following today. My own view agrees with that articulated by Prigogine: the basic processes we associate with dissipation and irreversibility—the linear nonequilibrium flows of mass, momentum, and energy which respond to gradients—deserve relatively simple explanations, local in time and space. Green and Kubo have already provided such simple explanations, by carrying Gibbs' approach one step further. Their linear-response theory of transport is the result. Green and Kubo's transport theory *is* irreversible.

The long-standing paradox that reversible equations of motion and irreversible behavior coexist can be addressed in a variety of ways. It is straightforward to apply perturbation theory to the statistical ensembles of Boltzmann and Gibbs, obtaining Green and Kubo's linear-response theory of transport. This approach is restricted to infinitesimal *linear* deviations from equilibrium. Boltzmann's equation applies arbitrarily far from equilibrium, but is restricted to low-density gases. Boltzmann "showed" that such systems invariably and irreversibly progress from less likely to more likely states, with the "state" concept loosely defined. His approximate "proof" was restricted to gases. With the system state described through the one-body distribution function f_1, and, with a statistical irreversible assumption for the collision law, the corresponding Boltzmann entropy $S_B = -H_B = -Nk\langle \ln f_1 \rangle$ obeys the H-Theorem, increasing monotonically to the equilibrium value provided that the system is isolated. Boltzmann certainly suggested[†] that similar ideas should apply to liquids and solids, but in quite vague terms.

Computational thermomechanics is able to treat liquids and solids as easily as gases, even far from equilibrium. The accuracy of this approach is primarily limited by computer time. As a result of computer simulations we can see, and surmount, the difficulties that frustrate an analytic approach like Boltzmann's to understanding transport processes in dense fluids and solids. In dense fluids both coordinates and momenta have distributions qualitatively different to the equilibrium product distribution. Outside the linear-response theory, there is no useful way to approximate the many-body distribution with a product of one-body functions. The interparticle spatial correlations are dominant at high density. We can estimate the substantial errors incurred by the nonequilibrium product approximation

---

[†]The final sentence of his 1877 paper, "Über die Beziehung zwischen dem zweiten Hauptsatze der mechanischen Wärmetheorie und der Wahrscheinlichkeitsrechnung respektive den Sätzen über das Wärmegleichgewicht", emphasizes the difficulty of this task.

by considering the simpler equilibrium case. At equilibrium, the "excess" nonideal part of the dense-fluid entropy follows from the Mayers' virial series:

$$\Delta S/Nk = [S(N,V,T) - S_{\text{ideal}}(N,V,T)]/Nk =$$

$$-\sum_{n>1}[B_n + (\partial B_n/\partial \ln T)_V)]\rho^{n-1}/(n-1),$$

where the $\{B_n(T)\}$ are "virial coefficients" giving the density expansion of the pressure:

$$PV/NkT = 1 + \sum_{n>1} B_n\rho^{n-1}.$$

For hard spheres the data displayed on page 83 show that this series applies throughout the fluid phase. The excess entropy decreases with increasing density, reaching a value near $\Delta S = -6Nk$ at the freezing point, in rough conformity to the behavior of real "simple liquids". This decreased entropy means that the phase-space volume for freezing spheres is smaller than that from an ideal-gas phase-space estimate by a factor of $e^{6N} \simeq 400^N$.

We see today that the isolated low-density systems which were necessarily Boltzmann's primary interest differ from those constrained by or driven at their boundaries in two important ways: (i) steadily-driven systems eventually come to occupy attractors which have zero measure relative to the equilibrium distribution; (ii) the entropy of such fractal states is not at all well-defined. Both these differences are consequences of the strong correlations built up by nonequilibrium steady states. Evidently the equilibrium distribution is not a promising start for describing nonequilibrium systems, except in the linear-response regime of Green, Kubo, and Onsager.

Today *algorithms* for fast computers have replaced *analysis* as our primary means of "understanding". The difficulty of calculating many-body collisions can be overcome with molecular dynamics. Distributions can be characterized and dissipation can be related to the Lyapunov spectrum. There are also related numerical methods for solving the Boltzmann equation. Today these numerical approaches have largely replaced the outmoded analytic approaches of an earlier century. I describe a *numerical* approach to Boltzmann's equation in the next Section.

8.3 Boltzmann's Equation Today

Boltzmann's kinetic theory achieved several goals: (i) it described the "approach to equilibrium" through the H Theorem; (ii) it provided a derivation for the resulting equilibrium (Maxwell-Boltzmann) velocity distribution, (iii) it provided verifiable estimates for the dependence of the low-density transport coefficients on temperature and interatomic forces, and (iv) it gave us a means of following far-from-equilibrium time evolutions of dilute gases. Today, computer simulation provides simpler alternative approaches to these same four goals, and without the restriction to dilute gases. Simulation makes it possible to follow the relaxation effects described by the H Theorem, to characterize the spatial and velocity distributions and the transport coefficients. In addition, kinetic theory relates these transport coefficients to the shifts of the Lyapunov exponents in far-from-equilibrium gases and to the equilibrium correlation functions emphasized by Green and Kubo's linear-response theory.

Computers have likewise revolutionized the study of dilute gases. An important computational development, largely due to Græme Bird, the "Direct Simulation Monte Carlo" method, provides the best means to solve the Boltzmann equation numerically, for discrete statistical samples of typical particles. Because the number of particles studied is finite, typically ranging from millions to billions, fluctuations which are absent in the original Boltzmann-equation approach complicate the "Monte Carlo" solutions. This many-particle method imposes a grid of cells on the system under study, with a typical cell containing about a dozen particles, each with its own time-dependent location and velocity.

Following an evolving gas in a *six*-dimensional phase-space is already too much for a serious numerical calculation, but a *three*-dimensional purely spatial grid *is* possible. Bird's approach eliminates the three μ-space dimensions corresponding to the velocities, making low-density gas simulations feasible. The system is divided into *spatial* cells with a size approximating the mean free path, and the cell structure is used to select collision partners. The numerical time evolution of the system combines two separate operations during each timestep Δt: (i) a "free-streaming motion", for the time Δt, during which *all* the particles move, without collisions, followed by (ii) a separate collisional update, modifying the velocities of *some* pairs of particles, those chosen for collision "during" the time interval Δt. The collision probability is assumed to be proportional to the relative speed. "Colliding" particles need not be closer together than the width of a computational cell. Thus the method is inaccurate for cells larger than the mean free path between collisions. The collision parameters for pairs of particles chosen for collision in each cell are determined statistically. In two dimensions, for a fixed relative velocity, the possible collisional outcomes are tabulated according to the "impact parameter" b shown in Figure 8.2. All values of b up to the range $\sigma > b$ of the chosen forces need to be included. The necessary table gives the "scattering angle" χ as a function of the impact parameter and the relative speed $|v|$. From it the post-collisional velocities can be computed.

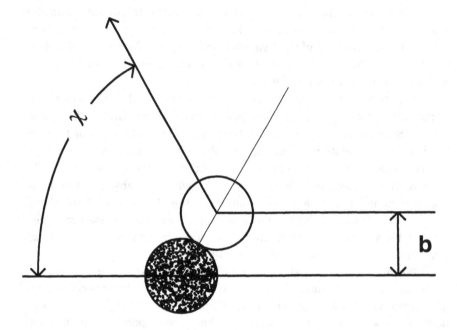

Figure 8.2: The closest distance of approach, in the absence of a collision, defines the *impact parameter b*. The change in relative velocity defines the *scattering angle* χ. The coordinates of the lefthand particle define the origin. In this coordinate system the motion corresponds to that of a "light" particle, with mass $m/2$, being scattered by an infinitely-massive particle at the origin. In this example the light particle moves from right to left.

In three dimensions all the possible pre-collisional orientations, for two colliding particles with a relative velocity v and an impact parameter less than σ, fill a *three*-dimensional cylinder of radius σ and length $|v|\Delta t$. In two dimensions, as shown in the Figure, these orientations occupy a strip of width 2σ. Thus in three dimensions an extra angle is required. In either case the probability for a pair of particles to collide is proportional to their relative velocity. The collision probability is *assumed* to be independent of the particle positions within the cell.

Because this statistical method of selecting collisions randomly, looking up the outcomes in a table, is so much simpler than following the exact N-body dynamics, the *number* of particles which can be treated exceeds that for conventional molecular dynamics by a factor of ten to one hundred. Direct Simulation Monte Carlo can provide convincing flow information, at the atomic level, with little programming effort. It is evident that both angular momentum conservation and the H Theorem are casualties of this numerical solution method. Additionally, fluctuations give rise to the same difficulties as in molecular dynamics.

8.4 Gibbs' Statistical Mechanics

As experience with statistical calculations was gained, based on Gibbs' systematic formulation of statistical mechanics, it became clear that the equilibrium thermodynamic entropy simply reflects the number of accessible microstates, $S_{\text{Thermo}} \equiv S_{\text{Gibbs}} \equiv -k\langle \ln f_N \rangle = k \ln \Omega_N$. In classical statistical mechanics, this result describes the entropy of liquids and solids, as well as gases, both dense and dilute. Gibbs' result expresses equilibrium thermodynamics in terms of the number of available microstates. An alternative approach to entropy is its direct thermodynamic evaluation by integration of the measured equation of state: $TdS = dE + PdV$. Molecular dynamics or Monte Carlo simulations can be used to make the required "measurements" of energy and pressure as functions of temperature and volume. The difficulties in applying Gibbs' statistical mechanics to real problems are *computational*, just as are the difficulties in applying Hamilton's motion equations.

Hamilton's motion equations point out the difficulty in extending statistical mechanics to nonequilibrium problems. Liouville's incompressible theorem, $\dot{f}_N \equiv 0$, shows that following the time-development of the fine-

grained probability density f_N—which *increases* over time in nonequilibrium steady states—using Hamilton's equations of motion can *never* give
any of the *nonequilibrium* entropy increases described by the Second Law of
Thermodynamics. With thermomechanics, and the realization that $\langle \dot{f}_N \rangle >$
0 corresponds to the rarity of nonequilibrium states, the use of Gibbs'
entropy away from equilibrium—a concept even Gibbs found puzzling—
became even less plausible. Chaotic dynamics furnishes a singular phase-
space description of nonequilibrium distribution functions which the traditional thermodynamic investigations had not even suggested.

For the N-body distribution function, Loschmidt and Zermélo's reversal
and recurrence objections to Boltzmann's proof of the H Theorem remain
as powerful as ever. Isolated systems can tend toward equilibrium only
briefly. Gibbs classic text *Elementary Principles in Statistical Mechanics*
begins with an echo of his 1884 talk, a careful exposition of Liouville's incompressible theorem, the basis for his microcanonical ensemble. To make
this base compatible with the Second Law Gibbs then had to introduce
the "coarse-graining" of Chapter 7. This picture requires spanning phase
space with (hyper)cubes of sidelength ϵ. Then Gibbs quite plausibly, and
correctly, argued that an initially compact localized phase-space density
would spread out and eventually contribute density to all accessible hyper-
cubes. This useful intuitive picture of phase-space mixing, though certainly
correct, does not pass muster at the *rigor mortis* level, as Krylov emphasized repeatedly in his book.

A more detailed and compelling rationale for Gibbs' coarse graining became apparent two generations later, with the recognition that Lyapunov
instability and chaos automatically provide the necessary mixing to validate
the ensemble viewpoint and to render the notion of a complete trajectory
illusory. Though curious, Gibbs remained leery of exploring in print the
details required to construct and describe nonequilibrium systems. Such
details necessarily include realistic boundary conditions and detailed descriptions at the atomic level, and so offer no convenient general analog to
Boltzmann's kinetic equation for dilute gases.

After the Second World War computer simulations began to be carried
out for many-body systems. In those cases using "realistic" force laws,
simulations validated Gibbs' simplification of equilibrium physics, showing
the equivalence of molecular dynamics' time averages and Monte Carlo's
phase averages. Accurate equilibrium thermodynamic properties could be
calculated either way. The agreement showed that there was no reason to

doubt the validity of Gibbs' statistical approach. For nonlinear nonequilibrium physics there is no simple analytic ensemble approach like Gibbs'. But his canonical distribution did lead naturally to the calculation of the *linear* transport coefficients using first-order perturbation theory. That "linear-response theory" or "Green-Kubo theory" computes the ensemble-averaged response (such as a shear stress or a heat flux) to an imposed macroscopic gradient (a transverse velocity gradient in the case of shear stress—a temperature gradient in the case of heat flux). The mass, momentum, and energy flows which result are all linear in the imposed gradients, providing theoretical bases for Fick's law of diffusion, Newtonian viscosity, and Fourier's heat conduction. Each of these nonequilibrium flows is "dissipative". The diffusive currents are eventually converted to heat. Without adding a thermostat to Gibbs' approach, to extract the heat generated by the perturbing gradient, there is no way to attain a true stationary state. An isolated system decays to equilibrium. A driven insulated system gradually gets hotter, as the inexorable heating continues. But because such irreversible heating is *quadratic* in the responsible gradient there is no special difficulty in obtaining the small-gradient linear transport coefficients for driven systems. Their calculated or simulated values can then be used in numerical macroscopic simulations of more-complicated nonequilibrium flows.

Of course, *linear* transport theory is not the full story. When materials are irreversibly compressed, shockwaves can result. Shockwaves are the consequence of the general increase of sound speed with increasing density. Compressive pressure waves steepen. The technical importance of shockwaves to understanding high-speed flight, blast waves, and chemical kinetics, has been responsible for a host of numerical studies. Some of them have been based on continuum mechanics. Some have been based on the Boltzmann equation. Both types of studies date back to the early days of computers. Shockwave simulations in dense fluids, using molecular dynamics, were described in Section 5.8. All such shockwaves are intrinsically irreversible nonlinear transitions—from an initial cold state, to a hotter, denser, higher-pressure state. The cold and hot system boundaries are relatively simple equilibrium states. This simplicity made possible the many shockwave simulations carried out during the past half century.

8.5   Jaynes' Information Theory

Any general approach to nonequilibrium states needs to handle both linear and nonlinear processes. Ed Jaynes embraced Gibbs' entropy as the basis for his own "information theory". Gibbs had already commented on the analogy between entropy and information in his boxed notes at the Yale University library. This *general* approach to constrained *non*equilibrium systems requires an extension of Gibbs' phase-space entropy from its natural isobaric isothermal equilibrium habitat, to far-from-equilibrium states. Jaynes suggested (i) maximizing the entropy subject to the restrictions that (ii) *all* essential nonequilibrium properties of the investigated system be included as constraints. For example, in addition to composition, energy, and volume, a homogeneous nonequilibrium system could additionally be constrained to have a specified stress component σ_{xy} or heat-flux-vector component Q_x. Such nonequilibrium requirements can be imposed by using Lagrange multipliers. The multipliers led formally to new generalized Gibbs-like ensembles with nonequilibrium probability densities varying exponentially in the additional constraints.

But we know now that true nonequilibrium distributions are infinitely more subtle than this. Imposing a single *stationary* constraint requires, in addition, Lagrange multipliers for *all* the time derivatives of that constraint (such as $\dot{\sigma}_{xy}, \ddot{\sigma}_{xy}, \ \dots$). Taking Jaynes' Lagrange-multiplier approach seriously would accordingly suggest adding *another exponential* controlling $\dot{\sigma}_{xy}$, and then *another*, constraining $\ddot{\sigma}_{xy}$, and so on. Presumably this infinite chain of restrictions implicit in Jaynes' work corresponds also to the practical translation of Zubarev's ideas for formulating nonequilibrium ensembles. Because purely Hamiltonian equations of motion do not satisfy any useful nonequilibrium constraints—such as constant shear stress or heat flux—it appears that Jaynes' or Zubarev's extensions of Gibbs' ideas to nonequilibrium flows are so cumbersome as to be fundamentally flawed. Rather than simply including thermostats, Jaynes' approach requires the evaluation of an infinite number of Lagrange multipliers.

I studied the simplest possible problem, thermostated two-body color conductivity, from Jaynes' point of view, in order to quantify the errors in the information-theory approach. At low density the thermostated two-body problem can be solved by assuming a statistical distribution of collisions, just as Boltzmann did. The stationary distribution of velocities which results depends on the direction of motion relative to the accelerating field direction. This distribution, $f(0 < \theta < \pi)$ is compared to the approximate one from Jaynes' maximum-entropy Lagrange-multiplier approach in Figure 8.3.

Figure 8.3: Comparison of velocity distributions f/f_0 for two-body low-density color conductivity. Jaynes' information theory (above) does not agree with the correct solution (below) from Boltzmann's equation for $f(\theta)$ where θ is measured relative to the field direction.

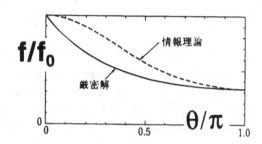

The hope that Jaynes' generalization of Gibbs' statistical approach *could* describe *non*equilibrium systems has a shaky basis. How could one reasonably expect to characterize nonequilibrium systems without ever describing the kinetic mechanisms necessary for maintaining them? *Dynamics is required* for an understanding of nonequilibrium systems. For many years the only significant progress made in this direction was, in my opinion, Green and Kubo's linear-response theory, a development important enough to warrant a Section of its own.

8.6 Green and Kubo's Linear Response Theory

The simplest version of Linear Response theory begins with Gibbs' canonical phase-space distribution for a system with Hamiltonian $\mathcal{H}_0$:

$$f_0 \propto e^{-\mathcal{H}_0/kT}.$$

In the event that the Hamiltonian is perturbed *slightly*, the corresponding linear change in Gibbs' canonical distribution function can be evaluated by formal perturbation theory:

$$\mathcal{H} \longrightarrow \mathcal{H}_0 + \Delta\mathcal{H} \; ; \; f \longrightarrow f_0 e^{-\Delta\mathcal{H}/kT} \simeq f_0[1 - (\Delta\mathcal{H}/kT)],$$

where terms of order $(\Delta \mathcal{H}/kT)^2$ are ignored. For example, the Doll's-Tensor Hamiltonian for periodic shear flow, which we mentioned in Section 3.9,

$$\mathcal{H}_{\text{shear}} = \mathcal{H}_0 + \dot{\epsilon}\sum y p_x,$$

provides equations of motion describing a flow field with the stream velocity

$$\langle \dot{x}(y) \rangle = \langle \partial \mathcal{H}/\partial p_x \rangle = \langle (p_x/m) + \dot{\epsilon}y \rangle = \dot{\epsilon}y,$$

and periodic shearing boundary conditions. The energy perturbation $\Delta \mathcal{H}$ induced by the strain rate acting for a particular time t is simply the stored energy, $\int_0^t \sigma_{xy}\dot{\epsilon}V dt'$. When the corresponding probability density perturbation, $\Delta f = -f_0(\Delta \mathcal{H}/kT)$, is used to calculate the mean shear stress,

$$\langle \sigma_{xy} \rangle = \int \cdots \int \prod(dq dp) \Delta f \sigma_{xy},$$

the shear viscosity coefficient $\eta \equiv \langle \sigma_{xy} \rangle / \dot{\epsilon}$, takes the form of a shear-stress correlation integral:

$$\eta = (V/kT) \int_0^\infty \langle \sigma_{xy}(0)\sigma_{xy}(t) \rangle dt.$$

In the small-strain-rate limit, the average can be computed numerically, using the *un*perturbed equilibrium distribution.

Perturbations which induce mass and energy currents similarly provide linear-response formulas for diffusion and heat conductivity. They also provide microscopic motion equations for the simulations of these nonequilibrium flows. These expressions link microscopic mechanics, properly ensemble-averaged, to the macroscopic phenomenological linear transport laws which characterize dissipative systems. This approach to the understanding of nonequilibrium systems begins with Gibbs' statistical mechanics and is perfectly general. It requires only an equilibrium time history. It is still true that any *practical* applications of Green-Kubo theory require molecular dynamics for a numerical evaluation of the transport coefficients.

The most important lesson from linear-response theory is that the irreversibility described by the Second Law requires *averaging*. From the pedagogical standpoint Green-Kubo theory is important in showing that Newtonian viscosity and Fourier's heat conduction are exact consequences of Gibbs' statistical mechanics. Specific nonequilibrium problems require more analysis. They require boundaries and thermostats incorporating the additional constraints and driving forces which distinguish nonequilibrium molecular dynamics—thermomechanics—from ordinary Newtonian mechanics. From the formal standpoint progress beyond linear-response theory requires confronting the fine-grained fractal information content which defeated Jaynes and Zubarev. Thermomechanics provides a practical means for discarding extraneous information through a computational coarse-graining, in which no more than fourteen digits need to be retained.

8.7 Thermomechanics

Steve Smale and David Ruelle suggested relatively early, though not so early as Krylov, that the classical mechanics of Newton, Lagrange, and Hamilton would have to be generalized so as to treat and discuss nonequilibrium systems in thermal contact with their surroundings. For engineers and experimental physicists, treating nonequilibrium surroundings as sources of feedback and control forces is routine. More theoretically-inclined physicists and mathematicians have been much less likely to include this useful approach in their calculations. Smale emphasized that exponential divergence of linear trajectory separations within a bounded space (such as an energy surface in phase space) must ultimately lead to a nonlinear bending and folding. The "Smale Horseshoe" idealizing this deformation embodies the idea of swirling mixing flow familiar from mixing paint.

Beginning with Fermi's numerical work at Los Alamos, which was continued there by Wood, and taken up by Alder and Wainwright at Livermore, simulation began to augment, and sometimes replace, the older theories of both equilibrium and nonequilibrium systems. In formulating algorithms suited to fast computers, it was necessary to develop computational analogs for thermal variables, like temperature. This could have been done indirectly, by fixing the energy of selected degrees of freedom constituting an *ergostat*, but the parallel with laboratory practice suggested instead *thermostats*, based on the ideal-gas temperature scale appropriate to classical mechanics. This idea led to modifications of the equations of motion in which the moments of the velocity distribution were controlled, either instantaneously, by using constraints, or on a time-averaged basis, by using feedback. New forces were added to Newton's equations of motion.

About 1970, computer simulations of driven steady-state systems, shear flows and heat flows, began to be carried out. These thermomechanical simulations of driven flows, using "nonequilibrium molecular dynamics", showed an overall agreement with phenomenological transport laws, reproducing viscosities and conductivities well despite relatively large fluctuations and uncertainties. The presence of gradients, generated with real boundaries, in nonequilibrium simulations, introduced a much larger number dependence than that associated with spatially-periodic equilibrium conditions. Though it was not explicitly stated in the earliest work, it soon was realized that these simulations used *time-reversible* motion equations and yet obtained *irreversible* behavior. The computer simulations revealed the same time-symmetry breaking and illustrated the same related paradox which had puzzled and stimulated Boltzmann, Loschmidt, and Zermélo— How *could* time-reversible motion equations yield irreversible behavior?

It is tempting to take seriously Loschmidt's argument that a time-reversible mechanics cannot lead to a future differing from the past. Stated this way, more recent investigations have shown that this point of view is naïve, though this is only known for sure in the case of time-reversible dissipative systems with phase-space compressibility. Such motion equations are *formally* reversible. Any solution of them can played back (with the coordinates appearing in reversed order, and with the momenta, and friction coefficients changed in sign). Although satisfying the same motion equations, all such played-back nonequilibrium solutions are unobservable and correspond to repellors. This is completely unlike the situation with classical Newtonian mechanics. So long as the boundary conditions are equilibrium ones, there are no known stationary Newtonian situations in which time symmetry is broken. Of course some transient Newtonian initial-value problems *are* effectively irreversible—the free expansion of a cold crystal, to form a gas, is irreversible because the initial conditions correspond to zero measure in the corresponding phase space. The confined free expansion discussed in Section 3.10.3 is an example of such an irreversible problem.

Generally, initial conditions cannot account for irreversibility. Zermélo and Loschmidt were right—apart from zero-measure initial conditions like the perfect cold crystal Newtonian dynamics cannot give long-time irreversibility. Green and Kubo's linear-response theory offers a more promising way out of the reversibility paradox. They found that the *average* response of equilibrium ensembles to small perturbations generating mass, momentum, and energy currents *can* be described by the diffusion, viscosity, and heat conductivity coefficients. It seems perfectly acceptable to take these results over to the continuum form of the conservation laws, from which the irreversibility described by the Second Law follows. The essential step is to replace time-reversible trajectory dynamics by an ensemble average of that dynamics along with a well-chosen perturbation.

8.8   Are Initial Conditions Relevant?

Another possible, but highly-artificial, way out of the reversibility paradox is to choose special initial conditions which *agree* with thermodynamic behavior, and to rule out any which do not. This approach seems to be a common refuge of philosophers who are unfamiliar with the insights furnished by computer simulation. The ultimate hard-to-prove modification of this reliance on initial conditions is to regress back to the Big Bang as *the* initial condition. This provides a low-entropy initial condition for the evolving parts of the Universe we have the luck and skill to observe. A second, more down-to-earth approach focusses instead on simple and idealized conservative systems.

Lawrence Sklar is a philosopher with a gift for explaining physics. Jean Bricmont is a staunch articulate defender of truth and simplicity in research. Both these men have argued that *the* initial conditions are important to any understanding of irreversibility. This point of view results from the desire to "understand" irreversibility by analyzing isolated systems obeying Newtonian or Hamiltonian dynamics. If one imagines that the hypothetical state of an isolated system is *exactly* known, then that state is *in principle* linked to others at all previous and all future times. This point of view suffers from the need to imagine an initial condition so precise as to contain the entire system trajectory for all previous and future times. This infinitely-detailed "view" of the system, including its entire history, parallels Price's "Block Universe", discussed below. I view the information content of such constructs as so unreasonable as to invalidate the entire argument. Given chaos, I see no way to reasonably study, or even contemplate, irreversibility based on the evolution of a *precisely-known* isolated system.

Huw Price is likewise a philosopher, and also feels that the "initial conditions" (of the universe!) are *the* crucial thing to understand. Price champions the view that time is not so different from space (where the positive and negative directions are arbitrary choices). Price adopts the "Block Universe" point of view, in which the "state of the universe" is laid out in both space and time, including all of space and all of time, both past and future.

With this task accomplished, he then sees no special difference between positive and negative directions, for either type of variable. He seems to miss entirely the necessity for defining, or accepting, a "subjective" time in order to *formulate* the underlying differential equations for a physical theory. Price believes that our quest for an understanding of dissipation and irreversibility is both futile and outmoded:

> *For all their intrinsic interest, the new methods of nonlinear dynamics do not throw new light on the asymmetry of thermodynamics. Writers who suggest otherwise have failed to grasp the real puzzle of thermodynamics—Why is entropy low in the past?—and to see that no symmetric theory could possibly yield the kind of conclusions they claim to draw.*

Price feels that the tendency of systems to approach equilibrium, the most likely state, is pretty well understood, and that the "Big Question" is how the Universe happened to start out in such an unlikely state. It seems to me that this "big question" is one of the "meaningless questions" (ill-posed questions, for which the path to an answer is inconceivable) about which Paul Schmidt warned me in my undergraduate philosophy class at Oberlin. It has become very clear to me, based on examples of the kind worked out in this book, that a simple understanding of irreversible processes *can* be based on the inherent relative stability of processes obeying the Second Law of Thermodynamics and the much greater *instability* of the time-reversed unobservable repellor versions of these same processes.

Thermodynamically irreversible processes involve the dissipation of energy and the growth of entropy. Continuum mechanics can incorporate a variety of processes, stationary or not. From the formal standpoint any process which conserves mass can be converted into a time-periodic process *if* it is possible for work to be done and heat to be extracted or added to the final state in such a way as to regain the initial state. Thus studying a time-periodic process is enough. In laboratory experiments it is an article of faith that the results depend only upon the *macroscopic* initial state, and not upon details at the atomic level. This same attitude has been adopted by those whose experimental medium is the fast computer. A clever choice of initial conditions can often reduce the time necessary to attaining a desired stationary state, but can never prevent it. The evolution algorithm, including boundary conditions, constraints, and driving forces, should be chosen so as to (i) cause the results *not* to depend upon the details of the initial state and to (ii) make analyses of the simulation as straightforward as possible.

The choice of boundary types is quite arbitrary. For example, a system can be contained within a boundary composed of solid or fluid particles interacting with relatively arbitrary force laws and subject to arbitrary mechanical and thermal constraints. Thermal constraints can be imposed by differential or integral feedback or by stochastic choices of velocities or accelerations. Among all these arbitrary choices the simplest forms of feedback lead to the simplest analyses.

Purely-theoretical analyses lack the efficiency, scope, and relevance of numerical steady-state simulation studies. *At* equilibrium, interesting fluctuations are quite rare. Choosing an initial state far from equilibrium results in early-time behavior dominated by the chosen initial conditions, and late-time behavior dominated by equilibrium fluctuations. It is arguably simpler, and certainly more educational, to follow the experimentalists, designing algorithms incorporating boundaries, constraints, and driving forces so as to study particular nonequilibrium states.

To me, the emphasis on the importance of initial conditions is doomed to failure for the two reasons cited by Loschmidt and Zermélo in their criticisms of Boltzmann's work on irreversibility. Eventually, an isolated system *will* return, and any forward-running trajectory *is* equally-likely to be found running backward. The presence of chaos also implies, as best articulated by Joseph Ford and Ilya Prigogine, that the very concept of *exact* initial conditions can have no operational significance. Thus, a meaningful discussion of physical correlations and evolution requires that the results *not* depend on all the unobservable details of the initial conditions.

Because the usual Hamiltonian systems are fundamentally inadequate for treating the very simplest nonequilibrium systems, steady states, and because the details of the initial conditions are quickly lost, by contraction below any observable scale, reliance on these initial conditions is a highly unproductive point of view. It needs to be realized that chaos makes computer simulation our *primary* source of knowledge for predicting the future. It is just as wrong to imagine that particular computer algorithms are the "cause" of the outcome as it is to ascribe macroscopic results to particular microscopic initial conditions. Both initial conditions *and* algorithms need to be irrelevant to the macroscopic outcome. The checks of *any* reasonable

computational approach are precisely those of laboratory experiments, reproducibility, intelligibility, and consonance with experience.

8.9   Constrained Hamiltonian Ensembles

Bob Dorfman, Pierre Gaspard, and Gregoire Nicolis have developed their
own novel boundary-based approach to "bridging the gap" between microscopic mechanics and continuum hydrodynamics. They consider the loss of
the mass, momentum, and energy escaping from a region within which the
usual conservative mechanics applies. For a sufficiently large region, the
long-time-averaged *escape rate* satisfies a macroscopic diffusion equation,
establishing a connection with standard macroscopic hydrodynamics. This
very simple physical idea is the basis of their "escape-rate theory". The
particle escaping from a finite Galton Board, in Figure 1.3, is a small-scale
example of this approach. Escape-rate theory leads to interesting mathematical conclusions. As time goes on, the particles remaining within the
Newtonian region dwindle, so that their phase-space states become more-
and-more specialized, eventually reducing to multifractal repellor distributions somewhat similar to those found in nonequilibrium steady states.

The escape-rate approach seems not to be intended to facilitate simulations or to attempt practical applications of nonequilibrium systems. It
is instead an intellectual and relatively formal attempt to understand the
physical meaning of the great (fractal) rarity of nonequilibrium states in
terms of standard Newtonian mechanics. It *is* significant that this approach
leads also to a fractal explanation of the dissipative states underlying the
Second Law of Thermodynamics. Otherwise, the complex style of the approach, and the lack of significant applications beyond Green-Kubo theory,
limit the appeal of the work. Escape-rate theory bears a family resemblance
to periodic-orbit theory. Rather than attack the problem of nonequilibrium
states headon, by adding and analyzing boundary conditions, driving forces,
and thermostats, an entire *equilibrium* ensemble is considered. Then *almost
all* of it is gradually discarded, bit by bit, until only the relevant fractal
part remains. This gradual pruning away of the equilibrium distribution
also characterizes Jaynes' approach, with an infinite number of Lagrange
multipliers, and the periodic-orbit approach, with an infinite number of
unstable orbits. In all these cases the desired states are those few remaining after all others, the vast majority, have been discarded. The pruning

process, corresponding to the loss of information is certainly an important part of irreversibility. But it is far better to allow the irrelevant information to depart quietly, through the negative Lyapunov exponents, rather than attempting the fruitless task of *analyzing* the details of that loss. Gödel guarantees that such a detailed analysis will ultimately have to fail.

8.10   Anosov Systems and Sinai-Ruelle-Bowen Measures

David Ruelle has discussed the possibility that the useful properties exhibited by certain oversimplified (and nearly vanishingly rare) dynamical systems, termed "Anosov systems", have counterparts in the more usual thermostated systems studied with nonequilibrium simulation methods. Anosov systems are oversimplifications, like square clouds or spherical chickens. The most "realistic" example I know of is the dynamics of a point mass constrained to travel a "surface of negative curvature", a surface, like the inner part of a doughnut, which *always* looks locally like a saddle. The defining property of Anosov systems is that they are "uniformly hyperbolic", with expanding and contracting saddle-like manifolds corresponding to their positive and negative Lyapunov exponents. Because nonequilibrium dissipative conditions correspond to an overall *negative* shift of the Lyapunov exponents the number of negative exponents typically exceeds the number of positive ones for large systems with a continuous spectrum. This rather common feature of time-reversible dissipation is evidently incompatible with the Anosov picture. In typical molecular dynamics simulations there is an incessant sign changing and reordering of the local Lyapunov exponents, associated with the wildly chaotic rotation of the "offset vectors" $\{\delta\}$ described in Section 7.4.

Although realistic differential equations cannot be expected to display the Anosov behavior, there are examples for maps, where discontinuities abet chaos. The Baker Map is the best illustration of consistent hyperbolic behavior, with a smoothly expanding unstable manifold and a stable contracting manifold along which the long-time-averaged distribution becomes fractal. Though it seems that there are actually *no* interesting smooth differential equations with the Anosov property, some mathematicians have found it stimulating to imagine and discuss them, with the avowed goal of gaining a better understanding of physics.

To date, there have been two "results" of this effort, both retrospective.

Both these theoretical results appear to me to have more to do with time reversibility than with the unrealistically-smooth and vanishingly-rare manifolds characterizing Anosov systems. The first result is a proof that "conjugate pairing" holds in nonequilibrium steady states. Conjugate pairing means that the sums of corresponding pairs of Lyapunov exponents—pairs which would always sum to zero at equilibrium—are likewise the same, for *all* pairs, away from equilibrium:

$$\{\lambda_1 + \lambda_N = \lambda_2 + \lambda_{N-1} = \lambda_3 + \lambda_{N-2} = \ ... \ = \lambda_{\frac{N}{2}} + \lambda_{\frac{N}{2}+1}\}.$$

So far as I know, such pairing results are invariably restricted to spatially homogeneous systems. Similar results hold also for non-Anosov, but homogeneous, Hamiltonian mechanics, with frictional damping added, *if* all the Cartesian momenta are damped with the *same* constant friction coefficient.

The second retrospective result obtained from the Anosov studies links the averaged time-integrated probabilities of likely and unlikely nonequilibrium states. In the case of a shear flow, for example, the probability ratio for time-averaged positive and negative shear stresses is predicted to be:

$$\text{prob}(+\sigma)_\tau / \text{prob}(-\sigma)_\tau = e^{+\Delta S/k} = e^{\int_0^\tau (\dot{S}/k)dt} \simeq e^{\tau \langle \dot{S} \rangle /k}.$$

The stresses are averaged over time intervals of length τ, as is also the external entropy change ΔS. Although the formula is reminiscent of the Einstein-Gibbs expression—used in Section 4.6 to estimate the probability of fluctuations away from equilibrium—the Anosov result can only be valid for sufficiently long times. As τ approaches zero and the dissipation integral vanishes, the predicted probabilities are equal.

Both the "results" just described— the Lyapunov-exponent pairing and the entropy-based flux probabilities—have been proven rigorously *for reversible Anosov systems*, despite the enduring lack of any interesting continuous examples of such systems. These "results" were actually given earlier by Denis Evans and several of his coworkers, for more general circumstances and through more elementary arguments. If there *were* such systems then it could be proved that they would generate relatively simple attractors, with equal numbers of positive and negative Lyapunov exponents. Because the simple geometric argument of Section 7.8 shows that nonequilibrium attractors are actually generated by *any* stable, time-reversible, steady dynamics, the applicability of the Anosov proofs is evidently rare to vanishing.

The oversimplified phase-space distribution associated with a dissipative Anosov system would evidently be somewhat intermediate between the volume-preserving flows of Hamiltonian mechanics and the attractor-repellor pairs of time-reversible nonequilibrium molecular dynamics. Sinai, Ruelle, and Bowen established the properties of such distributions and discussed the Lyapunov spectra for such hypothetical hyperbolic Anosov systems. The static hyperbolic structure gives a Lyapunov spectrum with a very simple structure, with fixed numbers of positive and negative local exponents. If the dynamics were also time-reversible then these two numbers would necessarily be equal. Analogs of the established relations were observed earlier in sufficiently simple computer simulations, typically homogeneous simulations with periodic boundary conditions.

Theoretical constructs, such as "measures", should be viewed with a healthy suspicion. The chaos inherent in interesting differential equations guarantees that our only access to the "strange sets" which constitute attractors and repellors will be representative time series from dynamical simulations. In no way can we construct, or even conceive of constructing, a Sinai-Ruelle-Bowen measure for an interesting system. On the other hand for very small attractors we *can* make rudimentary low-dimensional checks for ergodicity. For more interesting attractors, with dimensionalities of a few hundred, we can also obtain the entire Lyapunov spectrum and so estimate the attractor's Kaplan-Yorke dimension. Occam urged us to avoid the extraneous. Bridgman cautioned us to avoid adopting concepts which are not susceptible to an operational investigation. It is not surprising that Bridgman singled out set theory for special criticism.

More recently, Gallavotti and Cohen have emphasized the "nice" properties of Anosov systems. Rather than finding realistic Anosov examples, they have instead promoted their "Chaotic Hypothesis": *if* a system behaved "like" a [wildly-unphysical but well-understood] time-reversible Anosov system, there would be simple and appealing consequences, of exactly the kind mentioned above. Whether or not speculations concerning such hypothetical Anosov sytems are an aid or a hindrance to understanding seems to be an æsthetic question.

8.11 Trajectories *versus* Distribution Functions

Ilya Prigogine has devoted most of his lifelong research effort to supporting and promoting the study of "complex systems". Some 20 years ago he was a champion of the significance of the Baker Map. Today he makes strong claims for the merits of a formal approach which is nearly disconnected from the analysis of experiments or simulations. John Dougherty has kindly attempted to provide a perspective on Prigogine's work, critically comparing it to the information theory of Jaynes and Zubarev. See also Zetie's perceptive review. Prigogine has recently adopted the view that trajectories are themselves not a proper description of dynamics. His reasoning is quite consistent with Joseph Ford's longstanding observation—the information content of a chaotic trajectory is inaccessible in principle. This is because the information required to reverse such a trajectory grows in exact proportion to the trajectory's length. For this reason Prigogine chooses instead to emphasize the definition and evolution of distribution functions. He argues that distribution functions are a more faithful representation of the inherent uncertainty of physics.

Prigogine then claims to find that the time evolution of these distribution functions can be expressed in terms of two one-way "semigroup" structures. This gives a time-symmetry breaking analogous to the Second Law of Thermodynamics. Perhaps it is no coincidence that his point of view is not so different to that which comes directly from computer simulations. It is my guess that his symmetry-breaking "semigroup" approach corresponds to neglecting the unobservable repellors, and selecting the inevitable attractors, which latter are by now so familiar from simulation work with time-reversible thermostated systems.

The time-reversible deterministic computer experiments, beginning with the Galton Board study in 1986, showed that the trajectories *can* be followed stably in only one of the two time directions, *despite* time-reversibility. Further, the typical ergodicity of the computer simulations indicates that the trajectory and distribution-function approaches provide *identical* coarse-grained solutions. Prigogine's "rigged Hilbert spaces" for quantum systems probably correspond to the constrained fractal distributions which simulations generate for stationary nonequilibrium systems.

8.12   Are Maps Relevant?

Both the equilibrium and the dissipative Baker Maps were instructive aids to understanding in the early days of chaos. Elaborations of these maps continue to be studied today, particularly as models of spatially-extended systems. Chaotic ergodicity in maps requires only two variables, while differential equations require three, or possibly four, for a continuous situation. The dissipative but time-reversible Baker Map embodies many of the same paradoxical qualities as do its many-body dynamical analogs. The breaking of time symmetry is particularly transparent for maps.

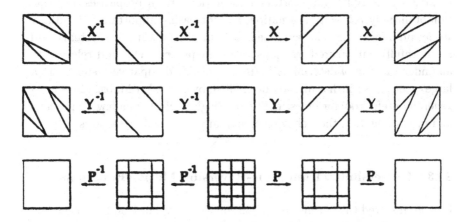

Figure 8.4: Time-reversible shear (X, Y) and reflection P maps in two space dimensions. Each of these three operations is "time-reversible" if the vertical y coordinate changes sign in the reversed mapping.

There is evidently an irresistable temptation to make maps more and more elaborate, to approach the complexity of continuum simulations by increasing the number and complexity of spatially distinct regions. *Time-reversible* maps must obey the defining relations:

$$M(x, y) = (x', y') \longleftrightarrow M(x', -y') = (x, -y),$$

where the "coordinate" x is unchanged by time reversal while the "momentum" y changes sign. Shears and reflections are the simplest examples of time-reversible maps. See Figure 8.4. Any symmetric combination of time-reversible maps, such as $M_{12321} \equiv M_1 \times M_2 \times M_3 \times M_2 \times M_1$ has also the time-reversibility property.

Maps tell us that symmetry breaking results in the net destruction of information, with an overall *decrease* in area corresponding to a concentration of probability density. The strange attractors which maps generate display a continuous *loss* of information. By using concatenated *chains* or *arrays* of maps, the shrinking can also diffuse to the interior of the chain as the result of boundary interactions. See the recent literature for work by Breymann, Tél, and Vollmer. The shrinking characterizing an attractor means that initially important information becomes gradually less so. At the same time the underlying chaos in a *strange* attractor generates expansion and the emphasis of initially unimportant information. Both properties, destruction of initially *relevant* information, and increasing importance of initially *irrelevant* information characterize nonequilibrium dissipative systems, and are faithfully caricatured by maps. Thus maps *are* useful and relevant to an understanding of deterministic time-reversible dissipative systems. To a large extent this rôle has already been filled, by the Baker Maps. Maps *are* a generally useful tool for investigating topological effects in a few dimensions and in investigating the behavior of spatially-extended aggregates.

8.13   Irreversibility from Time-Reversible Motion Equations

It is often stated that "time-reversible equations cannot lead to irreversible behavior". Though seemingly obvious, this statement is false. I learned this in 1986, by looking at Bill Moran's computer simulations of the isokinetic Galton Board problem described in Section 5.9.1. Independent of the detailed initial conditions, the Galton Board invariably generated the *same* fractal distributions and the *same irreversible* dissipative behavior, despite the *time reversibility* of the underlying equations of motion. Because the dynamics of the Board can be displayed in two dimensions, by cataloging collisions according to two characteristic angles, the evolution of this model, like the two-dimensional maps, is easy to visualize. The irreversibility exhibited by the Galton Board is this: invariably, for any initial conditions, stable chaotic flow states give rise to a current extracting energy from the field and dissipating it in the form of heat. The time-reversed flow, satisfying the same motion equations, corresponds to an unstable unobservable multifractal repellor, with heat converted into work. If one were to take the point of view that reversing time is simply replacing $+\Delta t$ by $-\Delta t$ in integrating the equations of motion then the dissipative flow just described would be occurring for decreasing time. But this artificial mathematical way of "reversing time" is unacceptable, for it discards the connection between velocities and coordinates, $\{\dot{r} \equiv v\}$, in favor of $\{\dot{r} \equiv -v\}$. The physicist accepts as basic an intuitive flow of time, its "subjective" or "psychological" "arrow of time", in distinguishing the past from the future and in making it possible to *write* differential equations. With this distinction in place, and with time flowing *forward*, he then can describe the changes resulting from time-reversible laws. *Dissipation* is the generic result, as is evident from the Galton Board example.

The dissipative Baker Map is an even simpler, though more idealized, example of the irreversible behavior hidden within time-reversible motion equations. Prigogine repeatedly emphasized the importance of the conservative version of this map in the early 1980s. Heat flow is a more physical example and involves no macroscopic mass currents. Analyses of heat conduction through a Newtonian region, sandwiched between two time-reversible heat reservoirs, one hot and one cold, provide a convincing demonstration of the loss of time symmetry through reversible dissipation. Such an example was considered in Section 5.9.3.

The *cause* of time-symmetry breaking is the systematic destruction of the information required to reverse a trajectory. Information corresponds to localization so that any contracting action in the phase space gives additional information. But, analogous to a coarse-grained cell in phase space, computers operate with a fixed wordlength, and have no way to accomodate *all* of this additional information. The loss of information which results becomes clear only gradually, for at any stage in a long trajectory, it is possible to reverse the trajectory reasonably accurately for several collision times. On the other hand, the distribution function obtained from the long-time-averaged trajectory is not symmetric in time, and invariably shows a net dissipative current parallel to the field direction. This current was already explained by Green and Kubo.

Because time reversal has no effect at all on the equations of motion, any attempt to find the past history of the system, at increasingly negative times, generates a "reversed" trajectory identical to the forward trajectory. When analyzed in the "forward" time direction, the reversed sequence of trajectories becomes a fractal repellor, differing from the attractor only in the signs of the velocities. This same symmetry applies to a heat transfer problem where the flow of heat in the "reversed" trajectory is from cold to hot, leading to a *negative-conductivity* repellor differing from the positive-conductivity attractor only in the sign of the heat flux component parallel to the temperature gradient. A reversed shear flow is slightly more complicated. In the formal time-reversed motion the strain rate changes sign while the shear stress does not. The quotient would give a *negative* shear viscosity, emphasizing the instability of the corresponding repellor. A time-reversed mass flow likewise corresponds to a negative diffusion coefficient.

8.14   Summary

It is evident that the difficult conceptual questions underlying the reversibility paradox and its resolutions cannot be usefully addressed in completely general terms. Examples are required. Despite this, the desire at least to understand, if not to control, the future, has led philosophers to ponder the meanings of time, probability, and irreversibility. Quantum mechanics provided philosophers with an unsettling need to consider probability. To me, it seems presumptuous of them to attempt a resolution of the argument between Bohr and Einstein, but they remind us that this subject is not yet closed. Maxwell recognized that Lyapunov instability destroys the apparent determinism of physics. Joseph Ford emphasized the same point. The impossibility of precisely specifying initial conditions for a chaotic problem seems lost on those who stress the importance of initial conditions. Any "good" computer simulation, meaning a *useful* algorithm, is certainly better controlled than most any laboratory experiment, and loses knowledge of the initial conditions in a time of order a few collision times. In computer simulations relevant to macrosopic physics, both the initial conditions and the chosen algorithm *must* be irrelevant to the macroscopic outcome.

A literary, completely theoretical approach to physics questions is unconvincing. Physics is grounded in observation with limited resources. Because computer simulations can be reliable sources of observation, they provide an essential adjunct to our understanding of physics. The exposure to philosophical ideas required for the preparation of this book led to the inclusion of many of the worked-out numerical examples. By providing tests of new ideas, examples can help to keep the literature relatively free of published errors[‡].

From the standpoint of physics it is quite clear that Boltzmann, followed by Green and Kubo, formulated the phenomenological transport laws governing mass, momentum, and energy flows as equilibrium ensemble averages. The necessity of averaging, ignoring the fluctuations, is obvious in all of this work. The result is a verification of the simplest phenomenological nonequilibrium constitutive relations. These can then be incorporated in the macroscopic simulations to provide results in fine agreement with experiment. They can also be used to extend the treatment of equilibrium thermodynamic systems to nonequilibrium systems, with a basis for both in microscopic mechanics.

It is interesting to contrast the sensitive chaotic instability of microscopic trajectories to the relentless predictable stability of macroscopic dissipation. The microscopic equations of motion generate wildly different solutions with only slightly different computers, or compilers, or computer languages. But the resulting transport coefficients and dissipative behavior are not at all sensitive to these effects.

The interventionist approaches necessary to nonequilibrium simulations were certainly foreseen by Ruelle and Smale. Krylov's agonized search for irreversibility could have ended with the discovery of linear-response theory, but there was then, and perhaps still *is*, more to do. The concept of a Lyapunov-unstable fractal is a key ingredient of our current understanding of reversibility. It was unknown in Boltzmann's day. The simple picture that flow proceeds from a terribly-unlikely past to an inevitable future is consistent with our everyday experience. To the extent that our knowledge of the current situation is incomplete we could explore the future of a corresponding ensemble, or of a typical initial condition. When the two provide similar answers we can conclude that a physical description is possible.

The mathematics of fractal sets remains as uncomfortably incomplete and paradoxical as it was in Bridgman's day. I heartily recommend a look at his 1934 "Second Look at Mengenlehre". How *can* the set of even integers be the "same size" as the set of all integers? How can it be that the number of points in an orange is equal to the number of points in *two* oranges? How can it be that there are fewer rationals than irrationals, while the binary grid composed of numbers with equal fractions of zeros and ones has exactly the same dimensionality as does the continuum (while any other fraction

[‡]For error references see my 1999 "The Statistical Thermodynamics of Steady States".

gives a lower dimensionality)? How can it be that *exactly the same points* make up *both* the inevitable attractor *and* the unobservable repellor?

Chapter 9

Afterword—a Research Perspective

Ah, make the most of what we yet may spend
Before we too into the dust descend;
Dust into dust, and under dust, to lie,
Sans wine, sans song, sans singer, and sans end.

Omar Khayyám, as translated by Edward FitzGerald

9.1   Introduction

In this final Chapter I take stock of our progress in defining and confronting
the reversibility paradox from the modern perspective of chaos and computer simulation. I summarize the way in which our current knowledge
resolves the long-standing problem of "understanding" irreversibility, and I
point out why some are still unsatisfied with this explanation. Our views,
and the challenges to them, change with passing time. I discuss my own
experience with innovation and change. Computers provide a mechanism
for change through the exploration of example problems. I stress the necessity and importance of well-chosen numerical examples to an understanding
of chaos and irreversibility. I also point out difficulties associated with an
overly-mathematical approach to these questions. Finally, I assess those
real problems, still in need of clarification, which are so nearly feasible now
as to constitute the research of the near-term future.

229

9.2   What do we know?

Boltzmann, Einstein, Gibbs, and Maxwell were all well aware that the Second Law of Thermodynamics holds only in an *averaged* sense. The statistical probabilistic point of view, so useful and familiar to us today, was forced on Boltzmann by his critics. By now we have added new knowledge, with the help of chaos and computers, to the integrated experience of the past. Nonequilibrium molecular dynamics (1972), Nosé-Hoover thermomechanics (1985), Moran's multifractal Galton Board simulations (1986), and the investigations of nonequilibrium Lyapunov spectra which followed these advances led us to our current state of knowledge. Today we see that even the smallest chaotic systems, when time-averaged, have behavior and properties fully consistent with the Second Law.

Chaos theory shows exactly *how* the macroscopic Second Law results from microscopic chaotic dissipative processes. Processes obeying the Second Law are stable, corresponding to attractors—they *destroy* the information which would be required to reverse them. This information flows *from* the system under investigation *to* its surroundings, where it is lost, as "heat". Processes violating the Second Law are unstable, corresponding to repellor states. Repellors are *only* observed formally, and artificially, by playing recorded attractor trajectories backward. They require an external source of relevant stored information. Chaos theory has led us from the blind alley of large conservative systems to the understanding of deterministic dissipation through thermomechanics. As forecast by many seers, this explanation of the Second Law *does* require an extension of mechanics to the maintenance of nonequilibrium states. It thereby avoids the silly emphases on dilute gases, "initial conditions", and isolated systems which have plagued this subject for so long.

Were it not for chaos initial conditions *would* be important. But chaos limits our descriptive abilities. Prigogine and Ford have correctly stated that chaotic trajectories are unobservable, beyond computation. The acceptance of chaos makes it possible to identify and exclude meaningless constructions, such as the precise trajectories of the Block Universe. Chaos makes it clear that irreversibility is not at all restricted to infinite systems. Irreversibility can be seen in very small samples, in long-time-averaged nonequilibrium steady states involving only one or two degrees of freedom. It follows that we must examine *sub*systems, "open systems" interacting with reservoirs, to understand real-world irreversibility.

Mechanical thermostats are essential to the simulation and understanding of open systems. Without them we could not so easily relate mechanics to thermodynamics and the real world around us. Time-reversible mechanical thermostats are a natural outgrowth of computer simulation. The time-reversibility property, pervasive in physics, turned out to simplify the implementation and interpretation of thermomechanics, and to lead naturally, through Liouville's Theorems, to an understanding of irreversibility.

At equilibrium, Gibbs' entropy $-k\langle \ln f_N \rangle$ and Shannon's "information" $\langle \ln f_N \rangle$ provide essentially the same picture. Maximizing entropy corresponds to minimal information. *At* equilibrium no understanding of dynamical processes or mechanisms is necessary. But, *away* from equilibrium the continual creation and destruction of information cannot balance. The imbalance is quantified by the information dimension of the strange attractors and by their Lyapunov instability. This is the lesson of the Second Law of Thermodynamics.

The inaccessibility and completely negligible measure of repellor states away from equilibrium, despite the formal reversibility of the equations of motion, makes this point clear in detail. Nonequilibrium core states occupy a vanishingly small fraction of the total states available at equilibrium. The difference between the embedding dimensionality of equilibrium states D_E and the nonequilibrium information dimension D_I reflects both the disappearance of phase volume and the divergence of the probability density f as the coarse-graining length ϵ is reduced toward zero:

$$\langle \otimes_{\text{noneq}} / \otimes_{\text{eq}} \rangle_\epsilon \propto \langle f_{\text{eq}} / f_{\text{noneq}} \rangle_\epsilon \propto \epsilon^{+\Delta D} = \epsilon^{(D_E - D_I)}.$$

This same relationship can alternatively be expressed in terms of the Lyapunov spectrum $\{\lambda\}$ and the external entropy dissipation rate $\dot{S}$. Kaplan and Yorke pointed out that the attractor information dimension can be estimated as the number of terms included in the partial sum $\lambda_1 + \lambda_2 + \ldots$ at the point where the sum changes sign, from positive to negative. With Nosé-Hoover dynamics the full sum is identically equal to $-\dot{S}_{\text{external}}/k$. Thus the nonequilibrium reduction in attractor dimension is approximately $\dot{S}_{\text{external}}/\lambda_1 k$, the number of terms which must be omitted to get a partial sum of zero:

$$\langle \otimes_{\text{noneq}} / \otimes_{\text{eq}} \rangle_\epsilon \propto \langle f_{\text{eq}} / f_{\text{noneq}} \rangle_\epsilon \simeq \epsilon^{(+\dot{S}_{\text{ext}}/\lambda_1 k)}.$$

Fractals are a universal feature of irreversible flows. That formally time-reversible nonequilibrium simulations invariably yield *fractals*, with detail at all scales, *was* a surprise. There have been many discoveries along the way. Ruelle had an early inkling of this, based on Anosov systems. Alder, Gass, and Wainwright's comparisons of Boltzmann and Enskog's kinetic-theory predictions with the results of molecular dynamics simulations, in 1970, emphasized the importance of chaos to irreversibility. Moran's work on the Galton Board showed fractals without any doubt. Prigogine found them in the Baker Map. Prigogine discussed semigroups, similar to the repellor-attractor pairs evident in the early computer studies. Much more recently Dorfman, Gaspard, and Nicolis showed how fractal structures arise when ordinary Hamiltonian systems are allowed to decay through escape at open-system boundaries.

It is wonderful to see now the profuse expression of Barnsley's idea [*Fractals Everywhere*, published by Academic press in 1988]. The fractal explanation of irreversibility is by now accepted and in common use, joining together the two ends of the spectrum connecting Boltzmann and Prigogine, and including all those many of us in between.

9.3   Why Reversibility is Still a Problem

Squaring nature's irreversibility with reversible equations of motion has been discussed, sometimes heatedly, for more than one hundred years. Why has this problem been with us so long? Partly, it is its difficult nature. We have no unified theory applicable to far-from-equilibrium systems. See Brad Holian's recent published lecture contrasting the "beauty" of presentday formalisms with the "ugly reality" of fractal distributions. Unlike equilibrium states, which are well-defined and unique once the state variables are given, the simplest *nonequilibrium* states—steady states—owe their existence and complexity to *boundary* conditions. More complicated states can also depend on time.

A common reaction to this intrinsic difficulty is denial. One can entirely ignore the need for boundaries, constraints, and driving, by choosing to study the "approach to equilibrium", beginning with special nonequilibrium initial conditions. Though this choice avoids a characterization of nonequilibrium states, it retains the useful equilibrium property of incompressible phase-space flows obeying standard Newtonian mechanics. This approach exaggerates the importance of initial conditions. It would baffle those concerned with real laboratory experiments where the irrelevant details of the initial state are never precisely known. Gibbs' equilibrium ensembles require no special boundary conditions and are time independent. At equilibrium the macroscopic properties require no time averages and no dynamical considerations. Nonequilibrium states are more difficult.

For how much longer will "irreversibility" be an interesting subject? The intrinsic interest of the subject guarantees continuing discussions of three kinds: (i) those intended to further the design and control of irreversible processes for practical needs, (ii) those intended to clarify the issue on intellectual grounds, and (iii) those intended to win fame and fortune for the participants. All three motivations will follow mankind into the unforeseeable future. The design and control of irreversibility for practical ends is in good shape despite the continuing need for clarification and explanation.

There is an understandable tendency for armchair philosophers to seek excessive generality. Their reach exceeds their grasp. Rather than confining their ruminations to well-posed problems, they prefer to visualize a "final solution", a "theory of everything". The urge to generalize can lead to oversimplification. It is startling to see discussions of physical theories which steadfastly ignore the fact that these *are* theories of *time evolution*. Change comes very slowly, reflecting Ford's favorite extract from Leo Tolstoy's work:

> *I know that most men, including those at ease with problems of the highest complexity, can seldom accept even the simplest and most obvious truth if it be such as would oblige them to admit the falsity of conclusions which they have delighted in explaining to colleagues, which they have proudly taught to others, and which they have woven, thread by thread, into the fabric of their lives.*

To what extent *is* Gibbs' ensemble viewpoint useful away from equilibrium? Because the phase-space probability density for simple nonequilibrium states is fractal, it is plain that the ensemble view is *not* very useful. The independence of results to initial conditions, for stationary or time-periodic nonequilibrium states, implies that the dynamics of a single system—Gibbs' time ensemble—provides just as much information as does the dynamics of an ensemble. So the utility of ensembles away from equilibrium must be decided on grounds of æsthetics or efficiency.

There certainly would be little more to prove if nature's underlying motion equations were themselves irreversible. Such a view seems totally unnecessary and is currently quite rare. Because there seems to be no compelling physical evidence for it, Occam would suggest avoiding this assumption of intrinsic irreversibility. For the more usual isolated mechanical systems, with many degrees of freedom and with phase volume conserved, the only irreversibility possible is the approximate type described by Boltzmann, the shedding of unlikely initial conditions. Isolated systems contain within themselves the paradoxical possibilities of return and reversal. With reversible nonequilibrium equations of motion, can there be irreversible behavior? The answer to this question is undeniably "yes", but for reasons which are difficult to understand, even for the simplest case, which is the reversible dissipative Baker Map.

Any study of irreversibility could be complicated further by introducing (somehow) gravitational, relativistic, and quantum effects. But these complexities seem totally unnecessary. To understand the dissipation associated with viscous and conducting flows, despite time reversibility, is today relatively simple. My view is this: the theoretical advances made since Boltzmann's discussion of dilute-gas irreversibility come from set theory, dynamical systems theory, an understanding of chaos, and the extension of Gibbs' ensemble theory to include linear response. A practical and believable understanding required, above all, digital computers and computer simulation. Beyond denial and dogmatism, a part of the reason for the long life of the reversibility problem lies in the complexity of *any* approximations or simplifications which could make possible a confrontation of the microscopic and macroscopic points of view beyond linear response theory: (i) one needs a clearly formulated mathematical theory, amenable to simulation and precise calculation, so that results can be obtained; (ii) a useful microscopic theory must necessarily contain precise, even if arbitrary, analogs of the relatively-vague macroscopic temperature and entropy; (iii) it must be admitted that numerical solutions, subject to Lyapunov instability, cannot attain precision for long times, so that a probabilistic description of the trajectory-based theory needs to be discussed. It is more difficult to pose meaningless questions and to pursue outmoded approaches if we keep in mind the need to achieve consistency among all three routes to knowledge: experiment, theory, and computer simulation.

9.4 Change and Innovation

Understanding irreversible flows has undergone qualitative change, from the smooth distributions of Maxwell, Boltzmann, and Gibbs to fractals. From simple analytic solutions of mechanical problems to the chaotic simulations of today. Boltzmann's work changed our views forever. He introduced probabilistic ideas. Despite the approximate and restricted nature of Boltzmann's gas-phase explanation of irreversibility, through the H Theorem, Lebowitz has promoted his view that Boltzmann's understanding of irreversibility was essentially complete, hampered only by an inability to work out phase-space integrals for condensed phases. This generous view credits Boltzmann with much more than he ever did, or would, claim. Boltzmann certainly did reach a nearly-complete understanding of the approach to equilibrium of dilute gases, but he could do little outside that area. It is certainly true that his fundamental relation $S = k \ln \Omega$, linking the *equilibrium* entropy to the number of phase-space states, is valid for *equilibrium* liquids and solids. His probabilistic collisonal assumptions, highly plausible (even correct!) *for gases*, provided a basis for the phenomenological irreversible macroscopic flow equations.

It is clear that Boltzmann had no idea at all about the existence of fractal distributions. Boltzmann certainly had no idea that simple differential equations of motion can lead to the complex irreversible fractal structures revealed by thermomechanical computer simulations. His accomplishments are monumental enough as is that it seems wholly inappropriate and unnecessary to credit him with more than what he actually did. The compressible Navier-Stokes equations, including Fourier's heat conduction as well as some interesting higher-order effects, and specialized modifications of the continuum equations to describe plasticity and fracture, are still quite adequate for engineers. They are not adequate for problems in which fluctuations are important, for which the modern numerical approach, Direct Simulation Monte Carlo, represents a distinct improvement over the analytic Boltzmann equation.

The passage of time leads inexorably to new concepts, tools, and vantage points from which past views seem limited, and of little value. Krylov's critical work, and the Ehrenfests' encyclopedia article, so valuable in its early summary of the accomplishments and the problems of kinetic theory, now hold little interest for a physicist, for whom Schrödinger's quantum mechanics, Maxwell and Boltzmann's kinetic theory, and Gibbs' statistical mechanics all coexist, though within slightly different areas of usefulness.

The creative and possessive impulses are active in every generation of researchers. Paraphrasing Niels Bohr: "Physics progresses one death at a time." Man has a deep-seated need for novelty, the replacing of old ideas with his own creations, which he is then loathe to relinquish. This natural evolution leads, over time, to the continual development of extremely esoteric methods for analyzing simple problems. Schrödinger even imagined concealing a microscopic source of randomness, modifying classical mechanics to introduce irreversibility. Prigogine has followed this lead for quantum systems. Systematic, highly-complex, barren formulations can be developed too. For example, a long trajectory, in a bounded space, necessarily nearly repeats, and so can be expressed in terms of periodic orbits in that space. Operators which advance distributions forward in time have eigenfunctions. These can be followed into the well-named complex plane.

Sometimes change is the result of an illusory quest for novelty. It is quite possible to pursue blind alleys in physics, roads through an imaginary landscape, which lead nowhere. I think that the Anosov systems, conjugate pairing theorems, and fluctuation theorems, represent progress only to the extent that they have analogs in the real world. So far there is no indication that something like pairing, or a fluctuation theorem, holds for a system with realistic nonequilibrium boundary conditions.

There is no doubt that trajectories and distributions propagate in different ways. A phase-space trajectory follows a completely-hypothetical one-dimensional track. It gains no complexity with passing time. Nevertheless, a Lyapunov unstable trajectory cannot be reversed by any means other than storing the results of the forward motion, and using these to direct the results of the reversed motion. The difficulties are even more severe in reversing a *non*equilibrium trajectory, for the sum of Lyapunov exponents is *positive* in the time-reversed motion. Any reasonable distribution spreads, due to Lyapunov instability, eventually filling out a coarse-grained probability density with no real clue as to the initial conditions. This disparity between trajectories and probability densities has led some to think the two approaches are profoundly different. Prigogine has recently asserted that trajectories themselves have no reality. Distributions *must* be considered.

Because distributions can be nothing more than the result of averaging a long trajectory this distinction is all the more puzzling to me. From the computational standpoint trajectories are obtained from *ordinary* differential equations. Summed-up coarse-grained trajectories then provide coarse-grained distributions. Trajectories are in fact the *only way* to get these distributions. Though formally these distributions satisfy a *partial* differential flow equation there is no practical way to implement that equation in interesting cases. The gradient operation ∇f is apparently ill-defined for typical fractals. In the end, the distinction between one-dimensional trajectories and the resulting fractal probability densities is certainly a red herring, as must be any explanation that distinguishes strongly (i) phenomenological reality, (ii) theoretical analysis, and (iii) computer simulation. There is no doubt that the three are separate pathways to a common partial understanding of reality and that including relativity, gravitation, quantum mechanics, or cosmology can enrich this understanding while adding nothing fundamentally new to our picture of irreversibility.

At the Livermore Radiation Laboratory of the early 1960s I soon discovered hurdles along the path of innovation. Francis Ree and I had to get Edward Teller's permission to use the computer time necessary for an accurate calculation of the hard-sphere solid entropy using constrained ensembles. Fortunately, Teller quickly agreed and that work was soon published. It wasn't until 1986 that I discovered publishing new ideas could be as difficult as generating and exploring them. When Bill Moran's solutions of the nonequilibrium Galton Board problem revealed multifractal objects, I was truly surprised. Though I knew what fractals were, from my work in Vienna with Harald Posch and Franz Vesely, I had not expected to see them in nonequilibrium simulations, and had even proposed to the National Science Foundation an investigation of nonequilibrium distribution functions using Fourier analysis! An editor at Physical Review Letters, George BasBas, rejected a manuscript describing our new findings. I sought him out, at a New York City cocktail party hosted by the Physical Society, in the naïve hope that he would have some interest in understanding the physics. But BasBas was steadfastly uninterested and relentlessly uninformed. Finally, by permuting the order of the authors' names and changing the title of the manuscript, we were able, on our third attempt, to publish it in Physical Review Letters. Meantime I had submitted another manuscript to the Journal of Statistical Physics, which was honoring Ilya Prigogine with a special issue devoted to the question of reversibility. Despite the relevance of my paper, Joel Lebowitz rejected it. Today I find it hard to believe that these early papers, which still seem to me to be clearly enough written, and certainly interesting, required extraordinary efforts to publish. But stories of this kind are legion. Jaynes' and Ruelle's are both well known.

Without the heros of the past we would have had no option but to reïnvent their discoveries. The luxury of building on the successful ideas from the past, and forgetting the rest, is what continues to provide the excitement and sense of progress still present today. From the long-term point of view the all too common priority disputes and struggles to differentiate minor results appear quite ludicrous. It seems increasingly rare today to find people willing to admit their mistakes. During my graduate student days at Ann Arbor things were different. When my advisor, Andy De Rocco, called Bob Zwanzig to report an error I had found, it took only a few seconds for Zwanzig to understand, and accept correction. Today even hours of conversation followed up with months of correspondence can be ineffective in resolving relatively simple questions. For references to some recent examples, work backward from my Physics Letters A manuscript of mid-1999.

9.5   Rôle of Examples

Examples are essential to progress on conceptual problems in physics. As it became apparent that fractal distributions underlie nonequilibrium problems occasional naysayers insisted that this view was mistaken. The simplest fractal examples are generated by maps. Maps and cellular automata replace the space-time continuum with discrete states. This simplifies counting states and can certainly return substantial conceptual rewards. The Baker map has both equilibrium and nonequilibrium forms. These forms can be concatenated into chains yielding large-system fractal distributions. Such a chain of maps is simpler and more interesting than Lorenz' continuous model, which can also be concatenated but which lacks time reversibility and has no equilibrium form.

With hindsight, it was possible to embed the successful thermomechanical simulation techniques into traditional mechanics. Whether or not the embedding is simpler than choosing *ad hoc* boundary, constraint, and driving forces is an æsthetic choice. The essential point is this: the perceived need to simulate the irreversible flows found in the laboratory led naturally to the time-reversible algorithms. These algorithms were only later linked to the structure of mechanics. Numerical solutions have provided valuable insights. Without them it would truly be impossible to improve upon Boltzmann's understanding. With them, the fields of Nonlinear Dynamics and Chaos have developed, providing new ideas, described with a rich vocabulary, accessible to physicists. The notions of mixing and Lyapunov instability coupled to Cantor's set theory are essential to a faithful description of the strange attractors accompanying dissipation.

The Galton Board is a particularly good model for the study of fractals and Lyapunov instability. It has both equilibrium and nonequilibrium versions. Its generalization to a many-body flow—the color conductivity problem—revealed the uniform shifts of Lyapunov exponent pairs called "conjugate pairing". Certainly it is worthwhile to complicate mechanics sufficiently to include the irreversibility of thermodynamics within it, as a consequence. This approach breathes life into Gibbs' ensemble theory, and allows us to apply some of his ideas away from equilibrium. Thermostats are also useful for *equilibrium* thermodynamics, for they provide mechanisms for generating Gibbs' ensembles from dynamical trajectories. Generalizing the concept of a mechanical system to a *thermomechanical* system, including reversible boundary interactions and constraints, provides a natural chaos-based irreversibility: the collapse to a relatively-stable strange attractor. The collapse, corresponding to information loss, or, equivalently, to the extreme rarity of nonequilibrium states, is the inevitable consequence of this approach. Beyond the reward of establishing a mechanism for irreversibility, the thermomechanical approach also avoids the completely hypothetical and unrealistic large-system limit. The *same* time-reversible dissipative behavior can be seen in small systems. The simple dissipative Baker Map and Galton Board are the prototypical examples, for discrete and continuous time, respectively. In these cases, as is typical, a repellor-attractor pair forms, linking two ergodic fractal objects in a way which breaks time symmetry through the dynamical sensitivity called chaos.

All the important features of interesting nonequilibrium systems can be seen in simple example problems. Lorenz' Attractor is mixing and chaotic. The Galton Board adds reversibility and ergodicity to these properties. The thermostated oscillator, with a temperature gradient, shows that phase-space compressibility is not always identical to the external entropy production. These examples are my favorites. Each scientist has his own. This is how men learn.

9.6 Rôle of Chaos and Fractals

It has taken quite a while to see that chaos, fractal geometry, and time-reversible ergodicity, with mixing, are all vital to an understanding of the reversibility paradox. Though Maxwell, Boltzmann, and Poincaré recognized the importance of "sensitivity to initial conditions", it remained to computer simulation to show the *need* for this property in order to obtain results *independent* of initial conditions.

The variety of theoretical approaches to irreversibility which have all developed since the seminal discussions at Howard Hanley's 1982 Boulder Conference* reminds me of the earlier profusion of *ad hoc* integral equations in the 1950s. Eventually, with the development of molecular dynamics and Monte Carlo simulations, the equilibrium "structure of simple liquids" became a solved problem. The need and the advocates for the obsolete approaches disappeared. Likewise, the fractal ergodic nature of time-reversible *non*equilibrium states is just now becoming well-known. We can anticipate that the periodic orbits, maps, Anosov systems, and other exotic tools for "understanding" what is now a solved problem will likewise gradually disappear.

Chaos makes it possible to access *all* the states while fractals quantify the global *rarity* of nonequilibrium states. It is not easy to find an instantaneous analog of this property, but perhaps the quest is not impossible. Computer simulation will continue to play its useful rôle, providing the answers to the new questions, just as it has for the old.

9.7 Rôle of Mathematics

The unreasonable applicability of mathematics to physics expresses two rôles—the ability of computation to simulate real phenomena, and to stimulate mathematical advances. From the viewpoint of a computational physicist, or even a skilled theorist like Feynman, the mathematics appears to lag well behind our understanding. It was with some difficulty, in 1986, that I came to understand the importance of time-reversible deterministic simulations, with nonequilibrium boundary conditions, to a fractal explanation of the Second Law of Thermodynamics. At that time Boltzmann's approach, understanding infinite systems, with Hamiltonian mechanics and without

---

*The Proceedings make up volume **118** of Physica A, which appeared in 1983.

boundaries, was the only approach under serious study by the mathematicians. The need for something new had been articulated, but *nothing* had been done.

What *is* the rôle of mathematics in continuing these advances? Perhaps it will widen their appeal and accelerate exploration and understanding? For some reason mathematics, while no doubt inspired by the computer results, is presently so restricted and complex as to be of little help in understanding or in making further progress. Quite often the mathematical treatment requires the acceptance of unrealistic assumptions, special unphysical cases, or nonuniform convergence. These difficulties suggest that the underlying phenomena are not being properly treated.

But the numerical results *can* be understood, and are not really *so* complicated. The existence of fractals is an obvious consequence for any flow which is simultaneously chaotic and dissipative. When the first computer-simulation results were obtained—fractal attractor-repellor pairs, shifted Lyapunov spectra, decreased dimensionality in phase space (as opposed to just decreased volume)—it was relatively difficult to publish them. Only about ten years later did most physicists working in the field accept some version of the new ideas, usually on the basis of their own specialized models, with more or less restrictive assumptions, and expressed in their own jargon. As I have stated in more than 100 seminar and conference talks, time reversibility, coupled with the requirement of boundedness, implies that only flows with long-time-averaged contraction can be observed. This *is* the symmetry breaking underlying the Second Law. "Explanations", based on Sinai-Ruelle-Bowen measures which lie completely outside the domain of operational definitions and practical techniques, are relatively specialized exercises for the experts. It certainly *would* be useful to have results for instantaneous nonequilibrium fluctuations, but this seems to be far beyond the capability of the various theoretical approaches.

The Second Law of Thermodynamics is perfectly general, and, when properly averaged, applies to all nonequilibrium systems. On the other hand, it seems to be necessary to examine each individual special case to make useful predictions of constitutive behavior. Are there any guidelines or general rules, analogous to Gibbs' ensembles, which would simplify the computational task of predicting nonequilibrium behavior? Is there any way to discover the fractal nature of a distribution function by examining only a small part of the system? These problems lead us to a consideration of the puzzles which still remain.

9.8   Remaining Puzzles

Taking stock, we have seen that chaos, when (i) constrained, (ii) dissipative, (iii) stationary, and (iv) time-reversible, guarantees not only the irrelevance of initial conditions, but also the destruction and loss of information required to reverse a dissipative trajectory. From the computational standpoint this understanding is convincing, complete, and verifiable. Simulation *is* a sufficiently faithful representation of irreversible phenomena. It explains our inability to recapture the past as a destruction of information, an automatic book burning built into chaotic dynamics.

Nonequilibrium thermostats have finally taken their rightful place in traditional classical mechanics as well. Gauss' Principle of Least Constraint can be used to derive the isokinetic or isoenergetic equations of motion for a driven system. Energy surfaces for Dettmann's Hamiltonian obey the Nosé-Hoover equations of motion. This derivation is not possible for situations, such as prototypical heat flow, involving more than one thermostat. Hamilton's Principle of Least Action, specially emphasized by Feynman, provided Dettmann, Morriss, and Choquard with another starting point for thermostats.

Perhaps there is no lasting need to seek a better understanding of the fundamental reversibility paradox than that available through an analysis of the dissipative Baker Map. More-elaborate chains of maps have been studied by Breymann, Gaspard, Tél, and Vollmer. This work is but the beginning few links in a logical chain coupling simple maps to complex numerical solutions of the complete partial differential equations of continuum mechanics. A uniform probability, subject to the action of the map, spreads out as an exact analog of a biased random walk. An initially uniform probability density approaches the multifractal distribution shown in Chapter 2, with the average value of the information entropy,

$$S_{info} \equiv -\sum p \ln p = -\langle \ln p \rangle,$$

where the $\{p\}$ are the bin probabilities, decreasing by $(1/3)\ln 2$ with each iteration of the map and ultimately diverging.

Iterating the same map "backward" (with the same initial condition, which is unchanged by time-reversal), provides a second fractal structure, the "repellor". Though the details obviously depend on the initial point, the analog of a Baker trajectory, generates, over time, exactly the same phase-space structure as does a repeated iteration of the distribution function. See Fox' paper. All the same features, the deterministic formation of fractal Lyapunov-unstable and time-reversible repellor-attractor pairs, are ubiquitous features of time-reversible nonequilibrium chaotic simulations of dissipative systems. Maps show us that this rich behavior simply represents correlated operations on one-dimensional digit strings. There appear to be problems simpler than maps, but perhaps conceptually just as fruitful, involving the manipulation of digit strings. Is there an analog of simple shear for digit strings? To what extent does the paradoxical coexistence of reversibility and irreversibility persist in simple models of this kind? It appears that a minimum of four dimensions are required for a simple continuous dissipative and time-reversible ergodic flow. See the paper by Harald Posch and me in the 1997 Physical Review E. Are there topological reasons for expecting increased complexity as the necessary dimensionality increases? Can we claim to "understand" nonequilibrium flows on the basis of two-dimensional maps?

It is evident that the mathematicians still have *much* more to learn and perhaps more to teach us. Feynman's description in his Princeton story, "A Different Box of Tools"[†] describes the frustration which can result when physicists request help from mathematicians. The idea of fixed numbers of positive and negative (local) Lyapunov exponents, seemingly essential to a "rigorous" Anosov-based analysis, contradicts results from even the very simplest of realistic models. The four time-reversible differential equations describing an ergodic harmonic oscillator in a temperature gradient, for example, give rise to only a single positive exponent, and two negative ones, even for small fields. The *local* exponents appear to be able to take on *all possible combinations* of signs, from $(-,-,-,-)$ to $(+,+,+,+)$.

Perhaps the paradox linking reversibility and irreversibility is more subtle than the usual distinction between rational and irrational numbers, while including the same fundamental mathematical difficulty, that which underlies the Banach-Tarski paradox. French illustrated that paradox by constructing a one-to-one mapping, carrying a single point set into *two* disjoint sets, congruent with the original. French constructed two oranges from one in this way. Such "paradoxical decompositions" and constructions have the same peculiar nature as do "steady" many-to-one phase-space flows with continuously shrinking distributions. It would be useful for a kind-hearted and knowledgable mathematician to express his understanding of

[†]*Surely You're Joking Mister Feynman* (Norton, New York, 1985).

such paradoxical behavior in terms more accessible and relevant to physicists. Bridgman was one of those who noticed and criticized the tailoring of Cantor's new clothes.

Because quantum mechanics, like classical mechanics, is a fundamental theory with the same time-reversibility properties, it suffers from the same recurrence and reversibility problems as does the classical approach. But there is an additional fly in the ointment: the Schrödinger Equation contains no forces, only energies, and this complicates the search for a nonequilibrium quantum mechanics. Chirikov has stressed that quantum mechanics, with its discrete spectrum of energy states, is analogous to computer simulations of classical problems in a discrete solution space. The analogy seems quite good. The typical level of precision in classical simulations, fourteen digits, is similar to quantum uncertainties on the scale of Planck's constant. The quantum situation seems more complex, in that adding constraints—through Lagrange multipliers—to the Schrödinger Equation can provide nonequilibrium stationary wave functions, but does not seem to lead to any chaos in the form of exponential divergence.

Does finite-precision arithmetic have any important effect on the properties obtained from classical simulations? The difference between having many very long periodic orbits, with fixed precision, and the idealized inaccessible ergodic chaotic orbit implied by classical continuum-based mechanics is reminiscent of the distinctions between quantum chaos (if any) and classical chaos.

How can we better characterize the multifractal distributions which arise in these nonequilibrium flows? The Lyapunov spectra seem to have less structure than do the frequency spectra of solid state physics. Despite this simplicity, approximate representations of fractals as sums of periodic orbits are cumbersome in the extreme. Are there some computational ways, both "interesting" and "practical", to characterize the different types of multifractal attractors? These representations should reflect the dramatic differences in visual appearance of the few simple two- and three-dimensional cases accessible with computer graphics. Despite their similar fractal dimensionality, there are undeniably big differences between puffy cumulus clouds and balled-up wads of paper.

Computational power continues to grow. In the near future we can look forward to much more detailed insights into time-reversible nonequilibrium systems. It should soon become possible to characterize the large-system limit of the Lyapunov spectrum for nonequilibrium hard-sphere systems and to characterize the corresponding eigenvectors. Likewise, faster machines will be able to determine the small-system fluctuations in Poincaré recurrence times, and improve the characterization of the multifractal phase-space distributions.

9.9 Summary

We have reached the end of our journey. My goal has been to explain, as simply and clearly as I can, how irreversibility results from time-reversible mechanical laws. To me it is enough to understand how velocity differences or temperature differences lead, irreversibly, to dissipation and to entropy increase. In the linear case the explanations of Green, Kubo, and Onsager are undoubtedly correct, and are in full agreement with the predictions of Gibbs' ensemble theory and Boltzmann's kinetic theory. Far from equilibrium the explanation is even simpler and more compelling, and makes use of concepts from chaos and fractal geometry. Chaos makes it possible to do much more. We have maps, attractors, and typical Lyapunov spectra for a wide variety of simple nonequilibrium problems. These provide understanding bridging the gap between mechanics and thermodynamics.

Consider again the four-chamber heat-flow example from Chapter 5, in which, for *any* initial condition, the time-averaged flow of heat corresponds to a positive heat conductivity (and so to a flow consistent with the Second Law of Thermodynamics). By accumulating the heat transfers to and from the "hot" and "cold" regions of the system, the entropy change of the hypothetical external heat reservoirs interacting with the two thermostated regions can be recorded as a function of time. See Figure 9.1, which shows a typical time development of the cumulative averages as well as their sum. The instantaneous oscillations about the well-characterized mean slopes are typically large and appear without warning. A first step for an interpretive nonequilibrium theory would be an understanding of the frequency and amplitude of these fluctuations. The overall slopes of the integrated heat transmitted to the reservoirs are proportional to the heat conductivity of the fluid confined by the two heat reservoirs. The entire system is fully time-reversible. If we had recorded *all* the particle coordinates and velocities, as functions of time, the entire many-body trajectory could have been played backward, and would have satisfied exactly the same (*time-reversible!*) motion equations. Thus, on a time-averaged basis, the change in external entropy, due to the fixed temperature difference, *is* positive in the forward direction of time and *would be* negative in the backward direction. Like Boltzmann's H Theorem, this *is* an example of irreversible behavior obtained from time-reversible motion equations. But the numerical algorithm applies to liquids and solids just as well as it does to dilute gases.

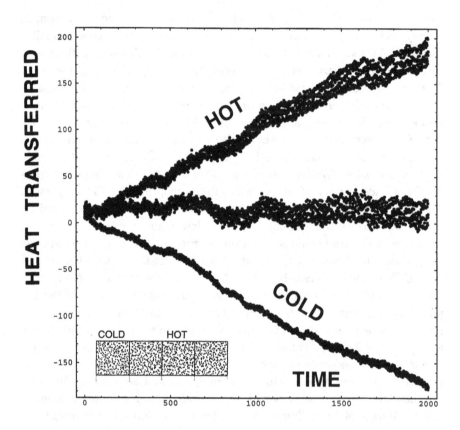

Figure 9.1: Cumulative dissipation in the four-chamber conductivity simulation discussed in Chapter 5. The total heat added to the hot chamber and that extracted from the cold chamber are shown. Their sum fluctuates in the neighborhood of zero, as energy balance implies.

But now consider the details of the Lyapunov spectra for the forward and backward trajectories. Forward in time the summed spectrum is *negative*, corresponding to the stable collapse to a multifractal attractor. The diaphanous nature of the attractor fits with our intuition that the number of nonequilibrium states is relatively small. If the trajectory, *and* its sensitivity to linear perturbations were *both* processed backward in time—a strict reversal of *all* the reference and satellite coordinates back toward their initial values—each of the Lyapunov exponents would change sign, both locally and as an average. Thus the summed spectrum of the reversed stored trajectory would have a *positive* sum and would correspond to a violation of the Second Law, even when time averaged. Why is such a reversed trajectory never seen? Precisely because of this instability. *Any* trajectory in the vicinity of the unstable repellor (corresponding to the reversed trajectory) promptly seeks out the stable attractor. For this reason the *formal* three-step construction of the past—(i) reversing the velocities, (ii) solving the equations of motion, and (iii) reversing the velocities *and* the time ordering of the points which result—is ineffective. It leads to an unobservable zero-measure repellor. It is quite clear that the stationary boundary conditions, because they extract information, lead to stationary heat flow which is only *formally* time-reversible. In fact the flow, together with its local Lyapunov spectrum, can *only* be reversed by playing the corresponding stored trajectories backward. This is the main message which fractals convey.

9.10 Acknowledgments

My interest in simulation was awakened during my college days, by the work of Berni Alder, Tom Wainwright, and Bill Wood. Andy De Rocco showed me how to use some of the tools of statistical mechanics, and encouraged my graduate studies. I was also strongly influenced by the clarity, depth, and sense of excitement in the kinetic-theory lectures given by Stuart Rice and George Uhlenbeck. After graduate school, Berni attracted me to the stimulating research atmosphere of the sputnik-era Radiation Laboratory at Livermore. There I met Steve Brush, who soon left for a distinguished career at Maryland, combining history with physics, and Francis Ree, who chose to remain at the Laboratory, taking advantage of the tremendous computational opportunities there, along with several other researchers interested in statistical mechanics.

A few years later, I was given the opportunity to teach statistical mechanics in the Livermore campus of the University of California, "Teller Tech". This University connection eased the pressure of the Laboratory, and also facilitated the travel so necessary to productive research. Among my students and researchers in the Department of Applied Science at Livermore, Bill Ashurst, Victor Castillo, Errol Craig, Oyeon Kum, Tony Ladd, Bill Moran, and Koichiro Shida contributed the most useful ideas. Koichiro spent several months at Livermore, doing research on Maxwell's "Thermal Creep" phenomenon, while preparing the Japanese translation of *Computational Statistical Mechanics* for publication. Kevin Boercker, Vic, Donovan Jones, and Nancy Owens provided help with the software required to illustrate this book. Son Xuan Nguyen carried out the scanning and sizing of the art work. Peter Raboin's Methods Development Group at the Lawrence Livermore Laboratory and the Academy of Applied Science (Concord, New Hampshire) furnished additional welcome support too. During my sabbatical leaves, Denis Evans, my son Nathan, Toshio Kawai, and Harald Posch were my main collaborators. Harald and Brad Holian have been constant sources of inspiration and new ideas ever since we first met.

Other colleagues, including Aurel Bulgac, Philippe Choquard, Christoph Dellago, Bob Dorfman, Marshall Fixman, Joseph Ford, Howard Hanley, Siegfried Hess, Gianni Iacucci, Dimitri Kusnezov, and Kris Wojciechowski have furnished stimulation and motivation from time to time. During my writing of this book Eddy Cohen, Dieter Flamm, Giovanni Gallavotti, and Harald were helpful in locating the relevant portions of Boltzmann's work. Dimitri Kusnezov and John Tully helped me in my efforts to understand Gibbs' views at New Haven. Brad, Carol, Christoph, Dimitri, and Oyeon kindly provided useful comments on early drafts of this book. Pierre Gaspard, David Ruelle, and Tamas Tél all offered generous comprehensive descriptions of their own points of view. Their views were helpful to me in articulating mine. Giovanni Ciccotti, Jean-Pierre Hansen, Siegfried, Sigeo Ihara, Karl Kratky, Michel Mareschal, Carl Moser, Shuichi Nosé, Bob Watts, and Kris facilitated a number of productive and instructive research tours for which I am grateful too. Finally, though last on my list, my wife Carol has been a constant source of the inspiration and happiness it has been my lot to enjoy while carrying out this work.

Glossary of Technical Terms

Algorithm: computational recipe suitable for computer evaluation

Attractor: long-time-averaged contracting *flow* solution, a fractal sink

Bit Reversibility: exact time reversibility, "to the very last bit"

Boltzmann Entropy: $-Nk\langle \ln f_1 \rangle$; f_1 is the one-body distribution function

Boundary Conditions: specified $v(r,t), T(r,t), \ldots$ on system boundary

Canonical Distribution: probability density proportional to $e^{-\mathcal{H}(q,p)/kT}$

Central Limit Theorem: N-fold averages from reasonable probability distributions follow a Gaussian distribution, with large-N halfwidth $\simeq N^{-1/2}$

Chaos: confined Lyapunov instability

Closed System: isolated system, with constant mass and energy

Coarse Graining: division of solution space into small cells

Conjugate Momentum: $p \equiv (\partial \mathcal{L}/\partial \dot{q})_q$, where q is a generalized coordinate

Conservative System: isolated system, with constant mass and energy

Constitutive Equation: state- (and sometimes rate-) dependence of fluxes

Constraint: specified value or history of thermomechanical variable

Continuity Equation: $\partial \rho/\partial t = -\nabla \cdot (\rho v) \longleftrightarrow d\ln \rho/dt = -\nabla \cdot v$

Continuum: system with continuously varying mass, momentum, stress, ...

Convergence: reaching a limit

Deterministic: time rate-of-change depends on current state

Diffusion Equation: $\partial \rho/\partial t = D\nabla^2 \rho$

Dissipation: conversion of mechanical energy into heat; heat loss

Dynamical Systems: coupled sets of ordinary differential equations

Energy: E, power to do work or create heat

Energy Equation: $\rho \dot{e} = \sigma : \nabla v - \nabla \cdot Q = -P : \nabla v - \nabla \cdot Q$

Ensemble: many similar systems with common specified state variables

Enthalpy: $H = E + PV$, thermodynamic potential for fixed pressure

Entropy: $S = \int (1/T) dQ_{\text{reversible}}$ integrated from reference to current state

Equation of Motion: differential equation for acceleration

Equation of State: relation linking pressure, energy, volume, temperature

Ergodicity: reaching all states consistent with macroscopic variables

Eulerian Coordinates: coordinates with respect to a fixed frame

Flux: quantity of flow per unit area and time

Fractal: power-law dependence of density distribution on (small) distance

Friction Coefficient: ζ, an inverse time; frictional force is $-\zeta p$

Gauss' Mechanics: $\{\dot{p} = F - \zeta p\}; \zeta = F \cdot p / 2K$

Generalized Coordinates: any variables sufficient to describe microstate

Gibbs' Entropy: $S = -k \langle \ln f_N \rangle$; f_N is the complete distribution function

Gibbs' Free Energy: $G = E + PV - TS$, thermodynamic potential for isobaric isothermal system

Hamiltonian: $\mathcal{H}(q, p) \longrightarrow (\dot{q}, \dot{p})$; usually an energy *sum*, $\mathcal{H} = K + \Phi$

Hausdorff Dimension: small-ϵ limit of $\langle \ln \# \rangle / \ln(1/\epsilon)$ (avoid this term)

Heat: energy in the form of velocity fluctuations

Heat Flux: comoving energy flow, per unit area and time

Helmholtz' Free Energy: $A = E - TS$, thermodynamic potential for isochoric isothermal system

H Theorem: Boltzmann's quantity $\langle \ln f_1 \rangle$ decreases monotonically for an isolated dilute gas (an approximation)

Ideal Gas: dilute gas with energy $E(T)$ and $PV \equiv NkT$

Information Dimension: small-ϵ limit of $\langle \ln \text{prob} \rangle / \ln \epsilon$

Invertibility: ability to recover reversed trajectory

Irreversibility: lack of time reversibility, usually dissipative

Lagrangian: $\mathcal{L}(q, \dot{q}) \longrightarrow (\dot{q}, \ddot{q})$; usually an energy *difference*, $\mathcal{L} = K - \Phi$

Lagrangian Coordinates: coordinates comoving with a moving material

Linear Response: ensemble-averaged response to infinitesimal perturbation

Liouville Equation: phase-space continuity equation for $\dot{f}(q, p)$

Liouville Theorem: $\dot{f} = -f \nabla \cdot v$, where usually $v \equiv \{\dot{q}, \dot{p}\}$ or $\{\dot{q}, \dot{p}, \dot{\zeta}\}$

Lyapunov Exponents: long-time-averaged orthogonal corotating and comoving growth (or decay) rates for infinitesimal perturbations

Lyapunov Instability: exponential growth of infinitesimal perturbations

Map: recipe for generating next configuration from present one

Mass Flux: mass flow, per unit area and time, ρv

Microcanonical Distribution: all accessible states of fixed energy

Mixing: long-time loss of correlation with initial condition

Multifractal: fractal distribution with spatially varying power law

Nosé-Hoover Mechanics: $\{\dot{p} = F - \zeta p\}; \dot{\zeta} \propto [(K/K_0) - 1]; K_0 \propto T$

Open System: system interacting with surroundings

Ordinary Differential Equations: expressions for *time* derivatives

Pair Potential: interaction between two bodies

Partial Differential Equations: expressions linking *space-time* derivatives

Periodic Boundary: opposite sides of system connected

Phase Space: (q, p) space, with points giving complete state description

Poincaré Section: surface, or plane, intersecting phase-space trajectories

Pressure Tensor: force per unit area—comoving momentum flux

Probability: likelihood, with $\sum \mathrm{prob} \equiv 1$

Random Numbers: uncorrelated, usually evenly distributed, with $0 < R < 1$

Repellor: time-reversed attractor, *source* of flow in remote past

Reservoir: external source of energy or momentum for system

Reversibility: same equations of motion in either time direction

Second Law: increase of entropy

Self Similar: same topological structure on smaller and smaller scales

Simulation: computer-generated solution of physical differential equations

Stability: indifference to small perturbations

Stationary State: boundary conditions fixed, see steady state

Steady State: boundary conditions fixed, see stationary state

Stochastic Boundary: boundary interaction including random numbers

Strain: relative deformation, $\epsilon_{xx} = \partial \delta x / \partial x$; $\epsilon_{xy} = \partial \delta x / \partial y + \partial \delta y / \partial x$

Stress: tensile force per unit area, $\{\sigma_{ij}\} \equiv -\{P_{ij}\}$

Temperature: comoving $T = \langle m v_x^2 / k \rangle = \langle p_x^2 / mk \rangle$

Thermodynamics: systematic study of energy transfers and conversions

Thermomechanics: mechanics with explicit heat reservoirs or fluxes

Thermostat: algorithmic mechanism for control of temperature

Virial Series: series expansion of the pressure in powers of density

Wave Equation: $\partial^2 u / \partial t^2 = c^2 \nabla^2 u$

Work: energy change due to the motion of macroscopic coordinates

Alphabetical Bibliography

[1] G. Benettin, L. Galgani, and J. M. Strelcyn, "Kolmogorov Entropy and Numerical Experiments", Physical Review A **14**, 2338-2345 (1976).

[2] L. Boltzmann, *Lectures on Gas Theory*, S. G. Brush, translator (University of California Press, 1964).

[3] J. Bricmont, "Science of Chaos or Chaos in Science?", Physicalia **17**, 159-212 (1995).

[4] P. W. Bridgman, "A Physicist's Second Reaction to Mengenlehre", Scripta Mathematica **2**, 101-117 and 224-234 (1934).

[5] P. W. Bridgman, "The Thermodynamics of Plastic Deformation and Generalized Entropy", Reviews of Modern Physics **22**, 56-63 (1950).

[6] V. M. Castillo and Wm. G. Hoover, "Heat Flux at the Transition from Harmonic to Chaotic Flow in Thermal Convection", Physical Review E **58**, 4016-4018 (1997).

[7] V. M. Castillo, Wm. G. Hoover, and C. G. Hoover, "Coexisting Attractors in Compressible Rayleigh-Bénard Flow", Physical Review E **55**, 5546-5550 (1997).

[8] C. Cercignani, *Ludwig Boltzmann, The Man Who Trusted Atoms* (Oxford University Press, 1998).

[9] N. I. Chernov, G. L. Eyink, J. L. Lebowitz, and Y. G. Sinai, "Derivation of Ohm's Law in a Deterministic Mechanical Model", Physical Review Letters **70**, 2209-2212 (1993).

[10] B. Chirikov, "Pseudochaos in Statistical Physics", 149-171 in *Nonlinear Dynamics, Chaotic and Complex Systems*, E. Infeld, R. Żelazny, and A. Galkowski, Editors (Cambridge University Press, 1997).

[11] P. Choquard, "Variational Principles for Thermostated Systems", Chaos **8**, 350-356 (1998).

[12] E. G. D. Cohen, "Boltzmann and Statistical Mechanics", 9-23 in *Boltzmann's Legacy 150 years after His Birth* (Lincei Academy, Rome, 1997).

[13] P. Coveney and R. Highfield, *The Arrow of Time* (Fawcett Columbine, New

York, 1985).

[14] L. A. Crotzer, G. A. Dilts, C. E. Knapp, K. D. Morris, R. P. Swift, and C. A. Wingate, "Sphinx Manual", Los Alamos National Laboratory Report LA-13436-M (April, 1998).

[15] M. Curd, "Popper on Boltzmann's Theory of the Direction of Time", 263-303 in *Ludwig Boltzmann Gesamtausgabe*, R. Sexl and J. Blackmore, Editors (Akademische Druck, Graz, 1982).

[16] Ch. Dellago, L. Glatz, and H. A. Posch, "Lyapunov Spectrum of the Driven Lorentz Gas", Physical Review E **52**, 4817-4826 (1995).

[17] C. P. Dettmann, "The Lorentz Gas: A Paradigm for Nonequilibrium Stationary States", in *Hard Ball Systems and the Lorentz Gas*, D. Szasz, Editor. To appear in the *Encyclopædia of Mathematical Sciences* (Springer-Verlag).

[18] C. P. Dettmann and G. P. Morriss, "Hamiltonian Formulation of the Gaussian Isokinetic Thermostat", Physical Review E **54**, 2495-2500 (1996), and references therein.

[19] J. R. Dorfman, *An Introduction to Chaos in Nonequilibrium Statistical Mechanics* (Cambridge University Press, 1999).

[20] J. P. Dougherty, "Foundation of Nonequilibrium Statistical Mechanics", 172-178 in *Nonlinear Dynamics, Chaotic and Complex Systems*, E. Infeld, R. Żelazny, and A. Galkowski, Editors (Cambridge University Press, 1997).

[21] P. and T. Ehrenfest, *The Conceptual Foundations of the Statistical Approach in Mechanics*, J. J. Moravcsik, translator (Cornell University Press, 1959).

[22] D. J. Evans, E. G. D. Cohen, and G. P. Morriss, "Probability of Second Law Violations in Shearing Steady States", Physical Review Letters **71**, 2401-2404 (1993).

[23] D. J. Evans and G. P. Morriss, *Statistical Mechanics of Nonequilibrium Liquids* (Academic, New York, 1990).

[24] J. D. Farmer, E. Ott, and J. A. Yorke, "The Dimension of Chaotic Attractors", Physica D **7**, 153-180 (1983).

[25] E. Fermi, J. R. Pasta, and S. M. Ulam, "Studies of Nonlinear Problems", originally a 1955 Los Alamos Laboratory Report LA-1940 and also available in the *Collected Works of Enrico Fermi* (University of Chicago Press, 1965). The 1972 article by Tuck and Menzel reviews this work and includes some interesting analysis of recurrence in nonlinear chains.

[26] R. P. Feynman, *The Character of Physical Law* (MIT Press, 1967).

[27] W. Fleischhacker and T. Schönfeld, Editors, *Pioneering Ideas for the Physical and Chemical Sciences: Josef Loschmidt's Contributions and Modern Developments in Structural Organic Chemistry, Atomistics, and Statistical Mechanics* (Plenum, New York, 1995).

[28] C. Foidl and P. Kasperkowitz, "Systematic Generation of Linear Graphs—Check and Extension of the List of Uhlenbeck and Ford", Journal of Computational Physics **89**, 246-250 (1990).

[29] J. Ford, "What is Chaos, that We Should be Mindful of It?", 348-372 in *The New Physics* (Cambridge University Press, 1989).

[30] R. F. Fox, "Entropy Evolution for the Baker Map", Chaos **8**, 462-465 (1998).

[31] R. M. French, "The Banach-Tarski Theorem", Mathematical Intelligencer **10**(4), 21-28 (1988).

[32] G. Gallavotti, *Statistical Mechanics* (Springer-Verlag, 1999).

[33] P. Gaspard, *Chaos, Scattering, and Statistical Mechanics* (Cambridge University Press, 1998).

[34] J. W. Gibbs, *Elementary Principles in Statistical Mechanics* (Oxbow Press, 1991). [First published in 1902.]

[35] E. Helfand, "Transport Coefficients from Dissipation in a Canonical Ensemble", Physical Review **119**, 1-9 (1960).

[36] B. L. Holian, "The Character of the Nonequilibrium Steady State: Beautiful Formalism Meets Ugly Reality", 791-822 in *Monte Carlo and Molecular Dynamics of Condensed Matter Systems*, K. Binder and G. P. F. Ciccotti, Editors (Italian Physical Society, Bologna, 1996).

[37] B. L. Holian, W. G. Hoover, B. Moran, and G. K. Straub, "Shockwave Structure *via* Nonequilibrium Molecular Dynamics and Navier-Stokes Continuum Mechanics", Physical Review A **22**, 2798-2808 (1987).

[38] B. L. Holian, W. G. Hoover, and H. A. Posch, "Resolution of Loschmidt's Paradox: the Origin of Irreversible Behavior in Reversible Atomistic Dynamics", Physical Review Letters **59**, 10-13 (1987).

[39] W. G. Hoover, "Temperature, Least Action, and Lagrangian Mechanics", Physics Letters A **204**, 133-135 (1995).

[40] W. G. Hoover and B. L. Holian, "Kinetic Moments Method for the Canonical Ensemble", Physics Letters A **211**, 253-257 (1996).

[41] W. G. Hoover, C. G. Hoover, and H. A. Posch, "Lyapunov Instability of Pendulums, Chains, and Strings", Physical Review A **41**, 2999-3004 (1990).

[42] W. G. Hoover and B. Moran, "Viscous Attractor for the Galton Board", Chaos **2**, 599-602 (1992).

[43] W. G. Hoover, T. G. Pierce, C. G. Hoover, J. O. Shugart, C. M. Stein, and A. L. Edwards, "Molecular Dynamics, Smooth Particle Applied Mechanics, and Irreversibility", Computers and Mathematics with Applications **28**, 155-174 (1994).

[44] W. G. Hoover, H. A. Posch, B. L. Holian, M. J. Gillan, M. Mareschal, and C. Massobrio, "Dissipative Irreversibility from Nosé's Reversible Mechanics", Molecular Simulation **1**, 79-86 (1987).

[45] W. G. Hoover and F. H. Ree, "Melting Transition and Communal Entropy for Hard Spheres", Journal of Chemical Physics **49**, 3609-3617 (1968).

[46] Wm. G. Hoover, *Computational Statistical Mechanics* (Elsevier, New York, 1991). [*Keisan Toukei Rikigaku*, Japanese translation by Koichiro Shida with the supervision of Susumu Kotake (Morikita Shupan, 1999).]

[47] Wm. G. Hoover, "Mécanique de Nonéquilibre à la Californienne", Physica A **240**, 1-11 (1997).

[48] Wm. G. Hoover, "Isomorphism Linking Smooth Particles and Embedded Atoms", Physica A **260**, 244-254 (1998).

[49] Wm. G. Hoover, "Time-Reversibility in Nonequilibrium Thermomechanics", Physica D **112**, 225-240 (1998).

[50] Wm. G. Hoover, "Liouville's Theorems, Gibbs' Entropy, and Multifractal Distributions for Nonequilibrium Steady States", Journal of Chemical Physics **109**, 4164-4170 (1998).

[51] Wm. G. Hoover, "The Statistical Thermodynamics of Steady States", Physics Letters A **255**, 37-41 (1999).

[52] Wm. G. Hoover, K. Boercker, and H. A. Posch, "Large-System Hydrodynamic Limit for Color Conductivity in Two Dimensions", Physical Review E **57**, 3911-3916 (1998).

[53] Wm. G. Hoover and H. A. Posch, "Shear Viscosity *via* Global Control of Spatiotemporal Chaos in Two-Dimensional Isoenergetic Dense Fluids", Physical Review E **51**, 273-279 (1995).

[54] Wm. G. Hoover and H. A. Posch, "Numerical Heat Conductivity in Smooth Particle Applied Mechanics", Physical Review E **54**, 5142-5145 (1996).

[55] Wm. G. Hoover, H. A. Posch, V. M. Castillo, and C. G. Hoover, "Computer Simulation of Irreversible Expansions *via* Molecular Dynamics, Smooth Particle Applied Mechanics, Eulerian, and Lagrangian Continuum Mechanics", Journal of Statistical Physics (Submitted, 1999).

[56] M. Ichiyanagi, "Conceptual Developments of Nonequilibrium Statistical Mechanics in the Early Days of Japan", Physics Reports **262**, 227-310 (1995).

[57] R. Illner and H. Neunzert, "The Concept of Irreversibility in the Kinetic Theory of Gases", Transport Theory and Statistical Physics **16**, 89-112 (1987).

[58] E. T. Jaynes, "Violation of Boltzmann's *H* Theorem in Real Gases", Physical Review A **4**, 747-750 (1971).

[59] R. Klages, K. Rateitschak, and G. Nicolis, "Thermostating by Deterministic Scattering", Physical Review Letters (Submitted, 1999).

[60] M. J. Klein, "The Physics of J. Willard Gibbs in His Time", 1-21 in *Proceedings of the Gibbs Symposium, Yale University, May 1989*, E. G. Caldi and G. D. Mostow, Editors (American Mathematical Society, 1989).

[61] N. S. Krylov, *Works on the Foundations of Statistical Physics*, A. B. Migdal, Y. G. Sinai, and Y. L. Zeeman, translators (Princeton University Press, 1979).

[62] O. Kum and W. G. Hoover, "Time-Reversible Continuum Mechanics", Journal of Statistical Physics **76**, 1075-1081 (1994).

[63] O. Kum, Wm. G. Hoover, and C. G. Hoover, "Temperature Maxima in Stable Two-Dimensional Shock Waves", Physical Review E **56**, 462-465 (1997).

[64] O. Kum, Wm. G. Hoover, and H. A. Posch, "Viscous Conducting Flows with Smooth-Particle Applied Mechanics", Physical Review E **52**, 4899-4908 (1995).

[65] D. Kusnezov, "From Chaos to Nonequilibrium Statistical Mechanics", Czechoslovak Journal of Physics **49**, 35-87 (1999).

[66] D. Kusnezov, "Quantum Lévy Processes and Fractional Kinetics", Physical

Review Letters **82**, 1136-1139 (1999).

[67] D. Kusnezov, A. Bulgac, and W. Bauer, "Canonical Ensembles from Chaos", Annals of Physics **204**, 155-185 (1990).

[68] J. Lebowitz, "Boltzmann's Entropy and Time's Arrow", Physics Today **46**, 32-38 (September, 1993). See also the Letters section in volume **47** (November, 1994).

[69] J. Lebowitz, "Microscopic Origins of Irreversible Behavior", Physica A **263**, 516-527 (1999).

[70] D. Levesque and L. Verlet, "Molecular Dynamics and Time Reversibility", Journal of Statistical Physics **72**, 519-537 (1993).

[71] E. N. Lorenz, "Deterministic Nonperiodic Flow", Journal of Atmospheric Science **20**, 130-141 (1963).

[72] M. Mareschal, Editor, *Microscopic Simulations of Complex Flows* (Plenum, New York, 1989).

[73] M. Mareschal, "Microscopic Simulations of Complex Flows", 317-392 in *Advances in Chemical Physics* **100** , I. Prigogine and S. A. Rice, Editors (John Wiley & Sons, New York, 1997).

[74] M. Mareschal and B. L. Holian, Editors, *Microscopic Simulations of Complex Hydrodynamic Phenomena* (Plenum, New York, 1992).

[75] M. Mareschal and E. Kestmont, "Experimental Evidence for Convective Rolls in Finite Two-Dimensional Molecular Models", Nature **329**, 427-428 (1987).

[76] J. J. Monaghan, "Smoothed Particle Hydrodynamics", Annual Review of Astronomy and Astrophysics **30**, 543-574 (1992).

[77] B. Moran, W. G. Hoover, and S. Bestiale, "Diffusion in a Periodic Lorentz Gas", Journal of Statistical Physics **48**, 709-726 (1987).

[78] R. Morris, *Time's Arrows* (Simon and Schuster, New York, 1985).

[79] S. Nosé, "Constant Temperature Molecular Dynamics Methods", Progress in Theoretical Physics Supplement **103**, 1-117 (1991).

[80] R. Penrose, *The Emperor's New Mind* (Penguin, New York, 1989).

[81] H. A. Posch and R. Hirschl, "Simulation of Billiards and Hard Body Fluids", in *Hard Ball Systems and the Lorentz Gas*, D. Szasz, Editor. To appear in the *Encyclopædia of Mathematical Sciences* (Springer-Verlag).

[82] H. A. Posch and W. G. Hoover, "Equilibrium and Nonequilibrium Lyapunov Spectra for Dense Fluids and Solids", Physical Review A **39**, 2175-2188 (1989).

[83] H. A. Posch and Wm. G. Hoover, "Time-Reversible Dissipative Attractors in Three and Four Phase-Space Dimensions", Physical Review E **55**, 6803-6810 (1997).

[84] H. Price, *Time's Arrow and Archimedes' Point* (Oxford University Press, New York, 1996).

[85] D. C. Rapaport, "Time-Dependent Patterns in Atomistically Simulated Convection", Physical Review A **43**, 7046-7048 (1991).

[86] D. Ruelle, *Chance and Chaos* (Princeton University Press, Princeton, 1991).

[87] D. Ruelle, "Smooth Dynamics and New Theoretical Ideas in Nonequilibrium

Statistical Mechanics", Journal of Statistical Physics **95**, 393-468 (1999).

[88] D. Ruelle, "Gaps and New Ideas in our Understanding of Nonequilibrium", Physica A **263**, 540-544 (1999).

[89] J. Schnack, "Molecular Dynamics Investigations on a Quantum System in a Thermostat", Physica A **259**, 49-58 (1998).

[90] M. Schröder, *Fractals, Chaos, Power Laws* (W. H. Freeman, New York, 1991).

[91] E. C. Schrödinger, *Science, Theory, and Man*, Chapter III, "Indeterminism in Physics" (Dover, New York, 1957).

[92] L. Sklar, *Physics and Chance; Philosophical Issues in the Foundations of Statistical Mechanics* (Cambridge University Press, New York, 1993).

[93] S. Smale, "On the Problem of Reviving the Ergodic Hypothesis of Boltzmann and Birkhoff", Proceedings of the New York Academy of Sciences **357**, 260-266 (1980).

[94] S. D. Stoddard and J. Ford, "Numerical Experiments on the Stochastic Behavior of a Lennard-Jones Gas System", Physical Review A **8**, 1504-1512 (1973).

[95] T. Tél, J. Vollmer, and W. Breymann, "Transient Chaos; the Origin of Transport in Driven Systems", Europhysics Letters **35**, 659-664 (1996).

[96] J. L. Tuck and M. T. Menzel, "The Superperiod of the Nonlinear Weighted String (Fermi-Pasta-Ulam) Problem", Advances in Mathematics **9**, 399-407 (1972).

[97] G. H. Vineyard, Cover Illustration, Journal of Applied Physics **30**, (August, 1959). The description on page 1322 credits J. G. Gibson, A. N. Goland, M. Milgram, and G. H. Vineyard with the underlying work.

[98] L. P. Wheeler, *Josiah Willard Gibbs. The History of a Great Mind* (Yale University Press, 1951).

[99] W. W. Wood, "Early History of Computer Simulations in Statistical Mechanics", 3-14 in *Proceedings of the International School of Physics "Enrico Fermi", Course 97, Molecular Dynamics Simulation of Statistical Mechanical Systems*, G. P. F. Ciccotti and W. G. Hoover, Editors (North-Holland, 1986).

[100] W. W. Wood, "On Some Additional Recollections, and the Absence Thereof, About the Early History of Computer Simulations in Statistical Mechanics", 908-911 in *Monte Carlo and Molecular Dynamics of Condensed Matter Systems*, K. Binder and G. P. F. Ciccotti, Editors (Italian Physical Society, Bologna, 1996).

[101] K. Zetie, "Time's Quantum Arrow Revisited", Contemporary Physics **39**, 393-395 (1998).

[102] D. N. Zubarev, *Nonequilibrium Statistical Thermodynamics* (New York Consultants, 1974).

[103] R. W. Zwanzig, "Time-Correlation Functions and Transport Coefficients in Statistical Mechanics", Annual Reviews of Physical Chemistry **16**, 67-102 (1965).

Index